Base Units, Prefixes and Estimation

Base Units

A system of base units has been established that is common to all physicists (and other scientists) around the world. This internationally recognised standardised system of units is known as **Système International,** or **SI** for short. The seven base units are shown in the table. All other units can be expressed in terms of these base units, e.g. frequency (Hz or s^{-1}), velocity ($m\,s^{-1}$), force (N or $kg\,m\,s^{-2}$) and density ($kg\,m^{-3}$).

Physical quantity	Base unit
Length	metre (m)
Time	second (s)
Mass	kilogram (kg)
Electric current	ampere (A)
Temperature	kelvin (K)
Amount of substance	mole (mol)
Luminous intensity	candela (cd)

Prefixes

Physical measurements extend from the very small (e.g. nuclear radius) to the very large (e.g. distances to galaxies). While these can be expressed in standard form, i.e. a number between 1 and 10 followed by a factor to show the power of 10 (e.g. 6.35×10^{-7} m), it is often more convenient to use standard prefixes instead. Their use is widespread and convenient, and it is important to recognise the most common ones as shown in the table. For example, the wavelength of red light (value indicated above) is better expressed as $\lambda = 635\,nm$ as this makes comparisons with other wavelengths in the visible region relatively straightforward and more convenient.

Index value	Name and symbol	Example
10^{12}	tera (T)	terawatt (TW)
10^{9}	giga (G)	gigapascal (GPa)
10^{6}	mega (M)	megajoule (MJ)
10^{3}	kilo (k)	kilogram (kg)
10^{1}	deca (d)	decimetre (dm)
10^{-2}	centi (c)	centimetre (cm)
10^{-3}	milli (m)	millilitre (ml)
10^{-6}	micro (µ)	microgram (µg)
10^{-9}	nano (n)	nanometre (nm)
10^{-12}	pico (p)	picosecond (ps)
10^{-15}	femto (f)	femtometre (fm)

It is important to be able to use these prefixes when converting between units. For example, $3.5\,GW = 3.5 \times 10^{9}$ W, and $2.72\,mm^{2} = 2.72 \times \left(10^{-3}\right)^{2} m^{2} = 2.72 \times 10^{-6}\,m^{2}$.

Non-standard Units

A number of units that are used in physics are non-standard. These units are more sensible to use for certain measurements and when making comparisons. For example, the **electronvolt** (eV) is a more convenient unit when discussing energy levels in atoms; 1 eV is defined as the energy required to accelerate an electron through a potential difference of 1 V. Therefore, 1 eV = charge on 1 electron $\times$ 1 volt $= 1.6 \times 10^{-19}$ J. An **astronomical unit** (AU) is the mean distance between the Earth and the Sun and it is more convenient to give distances within the solar system in units of AUs; 1 AU = 150 million kilometres or 1.50×10^{11} m. Outside the solar system, distances are commonly measured in **light years** (ly) or **parsecs** (pc); 1 ly $= 9.46 \times 10^{15}$ m and

1 pc = 3.26 ly. In radioactivity, the **curie** (Ci) is often used instead of the SI unit **becquerel** (Bq); 1 Ci = 3.7 GBq. It is important to be able to convert between these different units.

Significant Figures

The numerical result of a particular calculation does not determine the accuracy or precision of the final value; the result can only be given to the **least accurate** value of the numbers used in the calculation. In physics, most answers are usually given to 2 or 3 **significant figures** (s.f.). For example, the mass of the electron is 9.10938×10^{-31} kg but this is normally used in calculations to only 3 s.f., i.e. 9.11×10^{-31} kg, and this is the value given in the formula and data sheets. A frequent exception to this is the rest energy (in MeV) quoted for sub-atomic particles, which is given to 6 s.f. as the differences between them are usually very small. Although calculators provide 6 to 8 digits for all calculations, the answer should always be rounded to the same number of significant figures as the original data.

Orders of Magnitude

In physics, it is often appropriate to determine the value of a parameter only to within an **order of magnitude**, i.e. to the nearest power of 10. This is best achieved by rounding each value in the calculation to just 1 s.f. and cancelling down to simplify the final value. For example, the kinetic energy of an electron moving at say two-thirds the speed of light is given by:

$$E_k = \tfrac{1}{2} \times 9.11 \times 10^{-31} \times \left(\tfrac{2}{3} \times 3 \times 10^8\right)^2$$

$$\approx \tfrac{1}{2} \times 10^{-30} \times 4 \times 10^{16}$$

$$\approx 0.5 \times 4 \times 10^{-14} \approx 10^{-14} \text{ J}$$

Another example is roughly how many seconds there are in a year. This is $365 \times 24 \times 86\,400 \approx 400 \times 20 \times 90\,000 = 4 \times 10^2 \times 2 \times 10 \times 9 \times 10^4 \approx 10 \times 10^2 \times 10 \times 10^4 = 10^9$ s, or a billion seconds. Expressing results as an order of magnitude allows values to be compared more easily. For example, the size of an atom is $\sim 10^{-10}$ m whereas the size of a nucleus is $\sim 10^{-15}$ m; the diameter of the Earth is $\sim 10^4$ km whereas the diameter of the Sun is $\sim 10^6$ km.

Estimation

One of the crucial skills in physics is estimation, which involves making sensible **assumptions** about the parameters in a calculation. Estimation requires skill, judgement and a sound knowledge of the underlying physics involved. For example, the question posed might be how many grains of sand there are on a beach, what the mass of the Earth is or how many stars we can see at night. Such questions involve making sensible assumptions that can be justified, and they often lead to an order of magnitude calculation.

Forecasting

Alternatively, estimation may involve changing the value of a parameter and **forecasting** what the result would then be. This requires some understanding about **proportionality**. Proportionality is represented by the symbol '$\propto$'; a simple example is Hooke's Law, which states that the tension in a spring is directly proportional its extension, i.e. $T \propto \Delta L$. If we double the tension the extension is also doubled, and if the tension is halved the extension is also halved. However, sometimes the relationship between two quantities is more complex. For example, the frequency f of a note plucked on a stringed instrument is governed by the equation $f = \frac{1}{2L}\sqrt{\frac{T}{\mu}}$. If the tension T is doubled in the string then the frequency will increase by a factor of $\sqrt{2}$. However, if the length L of the string is doubled then $\frac{1}{L}$ is halved and so the frequency is also halved since $f \propto \frac{1}{L}$.

SUMMARY

- **Scientific units** are based on the Système International or SI system
- **SI base units** are metre, second, kilogram, ampere, kelvin, mole and candela; all other units and quantities are derived from the base units
- **Prefixes** are used with the SI system units to indicate a multiple or a sub-multiple of the quantity; they are written in front of the unit name or unit symbol
- **Non-standard units** are also in common use where a more convenient unit for comparison is needed, e.g. electronvolt, parsec
- The **numerical result of a calculation** does not determine the accuracy or precision of the final value; it can only be given to the least accurate value of the numbers used in the calculation
- In physics, the answers to calculations are usually only given to 2 or 3 significant figures
- An **order of magnitude** calculation gives the answer to the nearest power of 10
- **Estimation** is a technique that allows a sensible answer to be derived based on sensible assumptions that have a sound physical basis
- **Forecasting** is based on how one parameter behaves with respect to another and involves proportionality

QUICK TEST

1. Which of the following are known as base units? kilogram, hour, newton, newton-second, second, degree Celsius, volt, ampere, metre?

2. momentum = mass × velocity. Use the base units to show that the unit of momentum is equivalent to N s.

3. How many joules are there in 5 eV?

4. The distance between the Sun and the planet Jupiter is 7.785×10^{12} m. Express this distance to 3 s.f. in terms of (i) km (ii) AU (1 AU = 1.5×10^{11} m).

5. The distance across a nucleus is approximately 1.4×10^{-15} m. What is this in terms of (i) fm (ii) nm?

6. The wavelength of red light is 635 nm. Express this wavelength in terms of metres in standard form.

7. A thin piece of wire is 0.0024 cm in diameter. Express this diameter in standard form in terms of metres.

8. The potential energy of an object is given by $E_p = mgh$. Give an order of magnitude calculation of the potential energy of a car of mass 2.35×10^3 kg raised through a height of 9.3 m. Assume that $g = 9.81 \, \text{m s}^{-2}$.

9. Show that the formula $E_p = mgh$ has the energy unit of joules.

PRACTICE QUESTIONS

1. A beam of electrons is accelerated through a potential difference of 5 kV. The kinetic energy gained by one electron is given by the equation $E_k = qV$, where q is the charge on one electron and V is the potential difference. [$q = 1.60 \times 10^{-19}$ C]

   a) Calculate the kinetic energy gained by one electron in:

      (i) joules

      (ii) electronvolts.   **[2 marks]**

   b) If 3×10^{12} electrons are used in the beam, estimate the total energy gained by all of the electrons in the beam and give your answer in terms of mJ.   **[3 marks]**

   c) 20% of the electron beam is absorbed before reaching the detector. Calculate the amount of energy reaching the detector. Give your answer to 3 s.f.   **[2 marks]**

2. In an experiment involving the measurement of the Young modulus, E, of steel wire, the equation used is given by:
$$E = \frac{F/A}{e/l} = \frac{Fl}{Ae}$$
   where F is the force (N), e the extension (m), A the cross-sectional area (m²) and l the original length (m).

   a) The diameter of the steel wire is 2.3 mm. Calculate the cross-sectional area of the wire in m².   **[2 marks]**

   b) The value for the Young modulus for steel is 210 GPa. Write this value in:

      (i) standard form

      (ii) MPa.   **[2 marks]**

   c) If the steel wire is 4.5 m long and a force of 50 N is used, determine the extension of the steel wire in:

      (i) m

      (ii) mm.   **[3 marks]**

   d) What is the effect on the steel wire if:

      (i) the force is doubled (assuming A and l remain constant)

      (ii) the cross-sectional area is doubled (assuming the force and original length remain constant)?   **[3 marks]**

3. The resistance of a piece of wire is given by:
$$R = \frac{\rho l}{A}$$

   a) The symbol ρ represents the resistivity of a material. Show that the units of resistivity are Ωm.   **[2 marks]**

   b) If the diameter of the wire is doubled, what is the effect on the resistance?   **[3 marks]**

   c) If $\rho = 1.7 \times 10^{-8}$ Ωm, $l = 0.8$ m and $A = 3.2 \times 10^{-8}$ m², carry out an order of magnitude calculation for the value of the resistance.   **[2 marks]**

   d) Using the values above determine the value of the resistance giving your answer to 2 s.f.   **[1 mark]**

   e) If the material is now changed for one with a higher resistivity value, what would be the effect on the resistance?   **[1 mark]**

Physical Measurements

Instrument Uncertainties

In the majority of physics experiments, measurements are undertaken with a variety of instruments such as a digital or analogue ammeter to measure electrical current, a vernier caliper or micrometer to measure width and a thermometer to measure temperature. All of these instruments have an associated **uncertainty** or **precision**. The uncertainty of a reading, or range of readings, can never be smaller than the precision of the instrument. Most instruments have to be calibrated to give an absolute value at some defined fixed point. This means that the uncertainty in the device is just **±½ the smallest interval** used to carry out the measurement. For example, if a top-pan balance can measure to a **resolution** of 0.1 g then its **uncertainty** or **precision** is ±0.05 g; if a thermometer is marked in intervals of 0.1°C then the uncertainty is ±0.05°C; if an analogue ammeter can read to an interval of 0.1 A then its uncertainty is ±0.05 A. A metre rule and other length-measuring devices are an exception since they cannot be calibrated accurately. A metre rule with a resolution of 1 mm has an uncertainty of ±0.5 mm at both ends, hence a total uncertainty of twice this, i.e. ±1 mm or ±0.001 m. When using any form of digital or analogue instrumentation such as an electronic balance or a multimeter, always record the **full reading** on the instrument; for example, if a digital voltmeter reads 1.46 V then this value is the one that should be recorded. The resolution of the voltmeter is therefore 0.01 V and an uncertainty of ±0.005 V, which is generally rounded up to ±0.01 V. On multirange devices, always use the smallest range that still allows the maximum reading to be taken, as this range is likely to have the best resolution.

Measurement Errors

When instruments are used to determine a value, the result obtained has some inherent degree of **measurement error** and **measurement uncertainty**, as distinct from the uncertainty in the device. However, measurement error and measurement uncertainty are not the same. The **uncertainty** of a measured physical quantity is the spread of values within which the true value can be expected to lie; the **error** of a measured physical quantity refers to the difference between the measured value obtained and the 'true value'. There are two key sources of error in all measurements: **systematic** errors and **random** errors.

Systematic and Random Errors

A **systematic error**:
- is constant throughout a set of measurements
- is an error that affects all the measurements in the same way
- may make the measurements all too large or all too small
- can be difficult to spot
- may result from instruments that have been incorrectly calibrated or from a parallax error
- cannot be minimised by repeating measurements.

A **random error** (sometimes called a '**reading error**'):
- is present in every measurement
- is an error that varies between successive measurements
- affects a measurement in an unpredictable way
- is equally likely to be positive or negative
- can be minimised by repeating a set of measurements and calculating a mean value.

Measurement	Instrument	Resolution	Uncertainty or precision
Length	Metre rule	1 mm	±1 mm
	Vernier caliper	0.1 mm	±0.1 mm
	Micrometer	0.01 mm	±0.01 mm
Time	Manual stopwatch	0.01 s	±0.2 s (reaction time)
	Electronic timer	0.01 s	±0.005 s
Mass	Top-pan balance	0.01 g	±0.005 g or ±0.01 g
	Spring balance	1.0 g	±0.5 g
Temperature	Mercury thermometer	0.1°C	±0.05°C (or ±0.1°C)

The following straight-line graphs depict the differences between **systematic** and **random errors**. In these examples, the line of true values should pass through the origin. Straight-line graphs that would be expected to go through the origin should never be forced to do so, as this could prevent a systematic error from being identified.

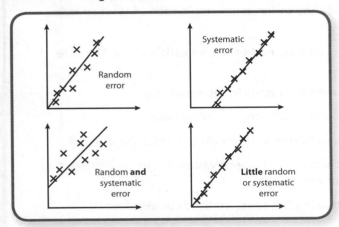

Accuracy and Precision

The **accuracy** of a measurement is a **qualitative** notion that tells you how close the measurements are to the 'true value' (whatever the 'true value' is). A result may be said to be accurate if it is relatively free from systematic errors. The **precision** of a measurement describes how close the measured values are to each other, and this may depend on the **sensitivity** of the instrument or device used (a sensitive device is one that responds with a large change in output for a small change in input). A precise measurement is one in which the random errors are small. Measurements can range from being precise and accurate to imprecise and inaccurate, as shown in the diagram.

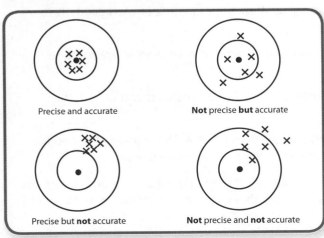

Precise and accurate   **Not** precise **but** accurate

Precise but **not** accurate   **Not** precise and **not** accurate

Reproducibility and Repeatability

Both of these terms describe how sets of results are consistent with each other. **Reproducibility** is when a set of results obtained using the same methods and techniques in different laboratories are consistent with each other to some degree of accuracy. The term **repeatability** is used to describe the ability to take further sets of measurements using the same method and within the same laboratory by the same person with a consistent degree of accuracy. Again, such measurements provide an indication as to the **reliability** of the results. In all experiments undertaken, the results should be both repeatable and reproducible.

Quoting Results along with Errors

- When giving results in terms of scientific notation or in standard form, always quote the value and the error with the same exponent
- Quote the result to the same number of significant figures as the quoted error implies
- Always quote the error to 1 or at most 2 significant figures

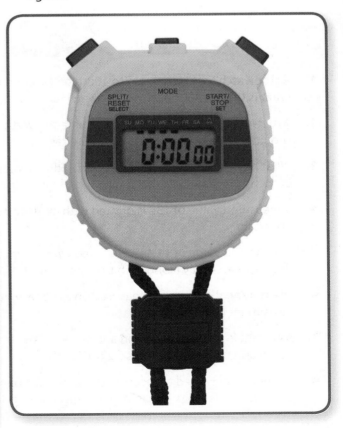

- The **resolution** of a device is the smallest non-zero interval that can be read on the device
- The **uncertainty** or **precision** of a **device** is half of the smallest division or interval marked on the instrument
- The **error** of a measured physical quantity refers to the difference between the measured value obtained and the 'true value'
- The **uncertainty** of a **measured** physical quantity is the spread of values within which the true value can be expected to lie
- A **systematic error** is an error that affects all the measurements in the same way
- A **random error** is an error that affects a measurement in an unpredictable way
- The **accuracy** of a measurement tells you how close the measurements are to the 'true value'
- The **precision** of a measurement describes how close the measured values are to each other; this depends on the ultimate sensitivity of the instruments used
- **Reproducibility** is when a set of results, obtained using the same methods in different laboratories, are consistent with each other
- **Repeatability** describes the consistency of further sets of results undertaken by the same method, within the same laboratory by the same person

1. A stopwatch that is accurate to 100th of a second is used to record timings of an object in motion. What is the resolution of the stopwatch and what would be a typical value for its precision?

2. A metre rule is being used to determine the vertical height of an object. Give two precautions that should be taken to ensure an accurate result.

3. What device can be used to measure widths typically less than a centimetre and what is the precision of such a device?

4. A measured value of 132 is quoted with an uncertainty of 18. Write the value to 2 s.f. along with the uncertainty.

5. A measured value of 11.448 is quoted with an uncertainty of 0.25. Write the value to an appropriate degree of accuracy along with the uncertainty.

6. The potential difference measured on a digital voltmeter is 3.36 V. Give this value together with the instrument uncertainty.

7. A current is measured with an analogue ammeter using a scale from 0 to 5 A. The reading obtained is 4.25 A and the interval size is 0.2 A. Give the value on the ammeter together with the uncertainty.

8. A metre ruler is used to measure the width of a bench. The ruler's smallest interval is 1 mm and the length of the bench is measured to be 65.4 cm. Express this length together with the uncertainty in metres.

considers both ends
of micr bw

PRACTICE QUESTIONS

1. A set of measurements for the diameter of a piece of wire is made and the results are shown in the table.

Diameter (mm)	5.01	4.94	4.98	4.92	4.95

   a) What is the name of the device used to measure such small distances? Give both the value of the resolution and the precision of this device.   **[3 marks]**

   b) What precautions should be taken before using this device?   **[1 mark]**

   c) The true value is 4.81 mm. By plotting the results or otherwise, explain whether the results are accurate and/or precise.   **[2 marks]**

2. A thermometer is used to record the temperature of water as it is heated from frozen. The results are shown in the table.

Temperature (°C)	1.0	2.4	3.9	5.0	6.3	7.1	7.9	9.6	10.1
Time (min)	0	1	2	3	4	5	6	7	8

   a) What is the resolution and uncertainty in the measuring device?   **[2 marks]**

   b) Draw a graph and plot the results of temperature (y-axis) against time (x-axis).   **[3 marks]**

   c) Draw the best line of fit through the points.   **[1 mark]**

   d) Given what you found in part **a)**, determine the nature of the results in terms of random and/or systematic errors and justify your conclusion.   **[4 marks]**

3. A steel rule is being used to measure the length of a metal bar that has a 'true' length of 795 mm. The rule can be read to the nearest millimetre. Repeated measurements give the following results.

Reading	1	2	3	4	5
Value (mm)	792	791	791	792	792

   a) What is the mean value for the length?   **[1 mark]**

   b) Are the readings accurate to 1 mm? Give a reason for your answer.   **[2 marks]**

   c) Are the readings precise to 1 mm? Give reasons for your answer.   **[2 marks]**

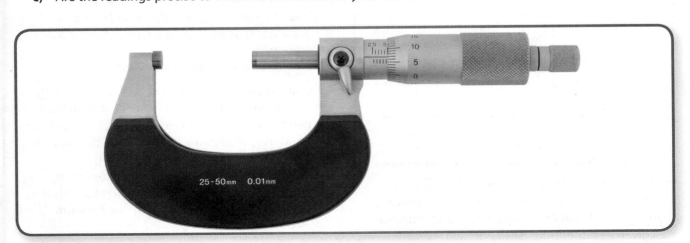

Errors and Uncertainties

Errors of Measurement

If just one value of a measurement is taken then this is fraught with uncertainties as there is no method available to ascertain its accuracy or precision (only the uncertainty associated with the measuring device itself can be determined). In order to define the uncertainty, accuracy and precision in a measurement, two or more values of the same parameter are required; it is then necessary to consider the errors in terms of differences from the mean value. For example, the potential difference recorded in a circuit gives values of 4.56V, 4.50V, 4.80V and 4.48V. The **resolution** of the instrument is 0.01V and the **uncertainty in the instrument** is ±0.005V; the **uncertainty in the measurement** is, however, much larger. The mean value of the reading is $V = 4.59$V, and the uncertainty in the data is found by:

measurement error, ΔV

$$= \frac{\text{range (largest value} - \text{smallest value)}}{2}$$

$$= \frac{4.80 - 4.48}{2} = 0.16$$

The **measurement error**, or **uncertainty** (in the measurement), is also known as the **absolute uncertainty** and is denoted below by Δa. All measurements should be expressed as:

mean value ± measurement error

$$a \pm \Delta a$$

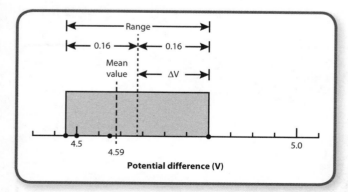

Potential difference (V)

In the above example, the potential difference is 4.59 ± 0.16V. The **uncertainty in a measurement** is therefore just ± **half the range**. This works remarkably well given its rather simplistic approach.

Absolute and Percentage Uncertainties

The above example allows a determination of the **fractional uncertainty** $\left(\frac{\Delta a}{a}\right)$. This is given by:

$$\text{fractional uncertainty} = \frac{\text{absolute uncertainty}}{\text{mean value}} = \frac{\Delta a}{a}$$

In the example given, the fractional uncertainty is:

$$\text{fractional uncertainty} = \frac{0.16}{4.59} = 0.035$$

and the **percentage uncertainty, $\varepsilon a = \frac{\Delta a}{a} \times 100$** or $0.035 \times 100 = 3.5\%$.

Combining Uncertainties

When **adding** or **subtracting** experimental values (say b and c), the **absolute uncertainties** are **added** together to give an overall experimental uncertainty in the value to be determined (i.e. a). In other words:
- If $a = b \pm c$ then the absolute uncertainty is given by $\Delta a = \Delta b + \Delta c$.

For example, in optics, if the object distance (u) is 5.0 ± 0.1 cm and the image distance (v) is 7.2 ± 0.1 cm then the absolute uncertainty in $(v - u)$ is ±0.2 cm or ±0.002 m and the final answer is given as 2.2 ± 0.2 cm or 0.022 ± 0.002 m [for comparison $(v + u)$ would be 12.2 ± 0.2 cm or 0.122 ± 0.002 m]. Note that in all calculations, including subtractions, the uncertainty always increases.

If two values (b and c) are **multiplied** together or **divided** then it is the **percentage** (or **fractional**) uncertainties that are **added** together to give the uncertainty in a:
- If $a = bc$ or $a = \frac{b}{c}$ then $\varepsilon a = \varepsilon b + \varepsilon c$.

For example, if the mass of an object is (30.2 ± 0.1) g and the volume of the object is (18.0 ± 0.5) cm^3 then the percentage uncertainty in the object's mass is 0.3% and the percentage uncertainty in the volume is 2.8%. The density is $\frac{\text{mass}}{\text{volume}} = \frac{30.2}{18.0} = 1.68$ g cm^{-3} and the percentage uncertainty in the density is $(2.8\% + 0.3\%) = 3.1\%$. Hence the absolute uncertainty in density is $1.68 \times \frac{3.1}{100} = \pm 0.05$ g cm^{-3} or ±50 kg m^{-3}.

Note, again, that the percentage uncertainty always increases when a calculation is made.

The density of the object is then given as $1.68 \pm 0.05\,\text{g cm}^{-3}$ or $1680 \pm 50\,\text{kg m}^{-3}$.

Finally, there is the **power rule**:

● If $a = b^c$ then the uncertainty is given by $\varepsilon a = c \times \varepsilon b$.

For example, if the radius of circle is $(6.0 \pm 0.1)\,\text{cm}$ then the percentage uncertainty in the radius is $\left(\frac{0.1}{6.0} \times 100\right) = 1.6\%$. The area of a circle is $\pi r^2 = \pi \times 6^2 = 113.1\,\text{cm}^2$ and the percentage uncertainty in the area is $2 \times 1.6\% = 3.2\%$. Hence the absolute uncertainty in the area is $113.1 \times \frac{3.2}{100} = 3.6\,\text{cm}^2$. The final area is given as $113.1 \pm 3.6\,\text{cm}^2$ or $110 \pm 4\,\text{cm}^2$. Note that the final answer has been rounded to 110, which is consistent with the number of significant figures in the absolute uncertainty. The uncertainty in any constant, including π, is usually taken to be zero.

These rules are summarised in the table.

Combination	Uncertainties
$a = b \pm c$	$\Delta a = \Delta b + \Delta c$
$a = bc$ or $a = \frac{b}{c}$	$\varepsilon a = \varepsilon b + \varepsilon c$
$a = b^c$	$\varepsilon a = c \times \varepsilon b$

Consistency between two Measurements

Two **independent measurements** of the same quantity can be said to be **consistent** if the two measurements $x_1 \pm \Delta x_1$ and $x_2 \pm \Delta x_2$ satisfy the following inequality:

$$|x_1 - x_2| < 3\sqrt{\Delta x_1^2 + \Delta x_2^2}$$

This means that the difference between the two readings can be explained by the uncertainty in the measurements. It can be said that they are consistent within experimental uncertainty.

SUMMARY

● An **error** is the difference between a measured value and the true value

● An **uncertainty** is a measure of the spread of values that includes the true value

● The **resolution** of a device is the smallest observable change in the measured quantity

● The **uncertainty in most measuring devices** is half of the smallest division or interval, except for digital meters where the value is generally rounded up to the smallest division

● An **uncertainty in a single measurement** depends on the uncertainty of the devices used

● An **uncertainty in a set of measurements** is given by $\pm$ half the range of the measured values; this is often called the **absolute uncertainty**, Δa. The measured value, a, is the mean of all the measured values

● The **fractional uncertainty** is defined as $\frac{\Delta a}{a}$ and the **percentage uncertainty** is defined as $\frac{\Delta a}{a} \times 100 = \varepsilon a$

● **Errors can be combined** depending on the form of the equation involving the quantity to be measured: adding or subtracting involves adding absolute errors; multiplying or dividing involves adding percentage errors

● The **consistency** between two independent measured values can be determined using the inequality $|x_1 - x_2| < 3\sqrt{\Delta x_1^2 + \Delta x_2^2}$

1. A thermometer is graduated in intervals of 1°C. What is the measurement uncertainty associated with this thermometer?

2. What is meant by the resolution of an instrument?

3. If the resolution of a set of weighing scales is said to be 0.1 g, what is the uncertainty in the values obtained?

4. An analogue ammeter is graduated in intervals of 0.2 A. What is the uncertainty of the device for recording current?

5. If the value of a measurement is a, what does Δa mean?

6. How is the percentage uncertainty determined from a single measurement whose value is a?

7. How is the absolute uncertainty determined from a range of measurements?

8. A particular resistor was measured on five occasions to give the following results: 1.20 kΩ, 1.16 Ω, 1.24 kΩ, 1.22 kΩ and 1.28 kΩ. What is the mean value of the resistor?

9. In the above set of results, what is the uncertainty associated with the measuring device used?

10. In question **8**, what is the absolute uncertainty in the measurement?

11. In question **10**, what is the percentage uncertainty in the measurement?

PRACTICE QUESTIONS

1. The resistance of a component is being measured. The potential difference across it is 8.2 ± 0.2 V and the current through it is 0.8 ± 0.1 A. The resistance, R, of any component is given by the equation $V = IR$, where V is the potential difference and I is the current.

   a) What is the value of the resistance of the component? **[2 marks]**

   b) Determine the percentage uncertainties in both the potential difference and current readings. **[4 marks]**

   c) From part **b)**, calculate (i) the total percentage uncertainty in R and (ii) the absolute uncertainty in R. **[3 marks]**

   d) Give the final value of the resistance together with its uncertainty. **[1 mark]**

2. The density of a piece of metal in the shape of a cube is being determined. The mass of the cube is measured to give the following results: 34.5 g, 34.2 g, 34.7 g, 34.9 g and 34.1 g.

   a) Calculate the mean mass of the metal cube. Give your answer to an appropriate number of significant figures. **[2 marks]**

   b) What is the uncertainty in the weighing scales used to determine the mass? **[1 mark]**

   c) Determine the absolute and percentage uncertainty in the above set of measurements and give your answer in the form: mass $\pm$ uncertainty in the mass. **[3 marks]**

   The dimension of the cube is 2.3 ± 0.01 cm for each side.

   d) Determine the volume of the cube and calculate the percentage and absolute uncertainty in the volume of the cube. **[5 marks]**

   e) Density is given by $\rho = \frac{mass}{volume}$. Calculate the absolute uncertainty in the density of the metal and give your final answer in units of $kg\,m^{-3}$. **[4 marks]**

3. Hooke's Law states that the extension of a spring is directly proportional to the load, i.e.

   $$F = kx$$

   where F is the load in N, x is the extension in m and k is a constant, known as the spring constant.

   a) If the spring extends by 4.6 mm when a load of 15 N is applied, determine the value of the spring constant in $N\,m^{-1}$. **[2 marks]**

   The uncertainty in the extension is ± 0.5 mm and the uncertainty in the force is ± 0.5 N.

   b) Calculate the percentage uncertainties in (i) the extension and (ii) the load. **[2 marks]**

   c) Determine the absolute uncertainty in the spring constant and write your answer as spring constant $\pm$ uncertainty. **[3 marks]**

   Other measurements taken using the same spring give a set of spring constants of values 3300, 3240, 3190 and 3140 $N\,m^{-1}$.

   d) Using the result in part **c)** together with the four other results above, determine the mean spring constant and the measured uncertainty in this set of results. **[3 marks]**

Graphs

Types of Graphs

In physics, graphs allow useful information about the relationships between quantities (variables) to be readily displayed using the **Cartesian coordinate** system. Here the x-axis represents the **independent variable** (i.e. what is controlled or deliberately changed in an experiment) and the y-axis the **dependent variable** (i.e. what is measured in an experiment). It is easier to calculate variables and recognise anomalies from a straight-line graph. Therefore, in most cases, a graph representing experimental results is reduced to a straight-line graph of the form $y = mx + c$, where the **gradient** or slope is m and the **intercept** along the y-axis is c. There are only a handful of fundamental graphs that you should be able to identify and any unfamiliar graphs presented should be manipulated into one of these forms.

Transformations

All of the relationships in the table can be plotted as a straight-line graph provided that the correct quantities are assigned to the x- and y-axes. For example, the equation for kinetic energy, E, in terms of the velocity, v, is given by $E = \frac{1}{2}mv^2$. This is a straight line if E (the dependent variable) is plotted as a function of v^2 (the independent variable). The resulting graph has a gradient of $\frac{1}{2}m$ and a y-axis intercept of 0. The equation of motion $v^2 = u^2 + 2as$ expresses the square of the final velocity v in terms of the initial velocity u, the acceleration a and the displacement s. The equation can be rearranged as $v^2 = 2as + u^2$, which allows a straight-line graph to be drawn if v^2 is plotted along the y-axis and s along the x-axis. In this case, the gradient is $2a$ and the intercept is u^2. The table provides a useful guide to the manipulation of other functions to form a straight line.

Plotting Straight-line Graphs

- Always label each axis with the quantities being plotted; include units
- Choose your scales appropriately to fill as much of the graph paper area as possible
- Plot the points by drawing a cross with a sharp pencil
- Draw the best line of fit through or between the points with care using a ruler; it does not have to go through the origin
- Give your graph a title

Determining the Gradient

It is often necessary to rearrange the equation to obtain it in the form $y = mx + c$. This may require a couple of steps of algebra. To determine the gradient or slope m of your straight line, draw the largest possible triangle and note the coordinates of the points connecting the straight line. The difference in the y-values and the difference in the x-values allow the gradient to be determined by:

$$m = \frac{\Delta y}{\Delta x}$$

In almost all cases, the gradient has units.

For non-linear graphs, the gradient at any point can be determined by drawing a tangent to the curve at that point and constructing a right-angle triangle.

The gradient is then determined from the triangle by calculating the differences in the x and y values and using the above expression.

	Parabolic	Reciprocal	Inverse square	Exponential	Square root
Function	$y = kx^2$	$y = \dfrac{k}{x}$	$y = \dfrac{k}{x^2}$	$y = e^{kx}$	$y = k\sqrt{x}$
Example	$E = \dfrac{1}{2}mv^2$	$R = \dfrac{\rho L}{A}$	$F = G\dfrac{Mm}{r^2}$	$N = N_0 e^{-\lambda t}$	$f = \dfrac{1}{2\pi}\sqrt{\dfrac{T}{\mu}}$
Independent variable (x-axis)	v^2	$\dfrac{1}{A}$	$\dfrac{1}{r^2}$	t (plotted against lnN)	$\sqrt{T}$
Gradient	$\dfrac{1}{2}m$	ρL	GMm	$-\lambda$	$\dfrac{1}{2\pi\sqrt{\mu}}$

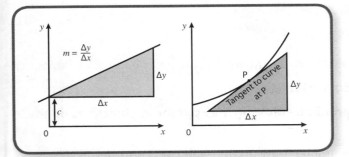

Error in the Gradient and Intercept

The absolute and percentage error in the gradient can be determined from drawing two more lines of fit through the points; the steepest line and the shallowest line drawn through the points provide a measure of the degree of uncertainty in the line of best fit. These worst-case lines are best drawn after error bars have been drawn. The line that shows the greatest discrepancy from the best line of fit is then used to estimate the absolute error in the line of best fit. Extending the worst-case lines to cross the y-axis also allows the absolute error in the value of the y-intercept to be assessed.

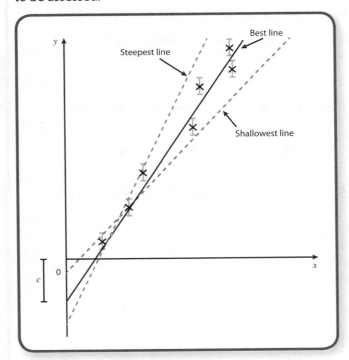

Error bars (or range bars) are plotted by drawing $\pm 0.5 \times$ range above and below the mean value.

Areas under Graphs

For some graphs, the area under the curve or line provides some specific physical meaning. Examples of this include:

- Area under a **velocity–time** graph gives **displacement**
- Area under a **force–extension** graph gives the **elastic potential energy** or work done/energy stored
- Area under a **potential difference–charge** graph gives the amount of **energy stored** in a capacitor

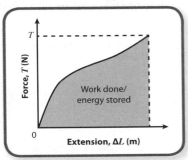

For straight-line graphs, the area can be calculated using the area of a triangle. For non-linear graphs, it may be necessary to count the number of squares under the curve and multiply this by the value of one square in the appropriate units. The units of the area can be calculated by multiplying the units of each axis.

SUMMARY

- Cartesian graphs **involve plotting** y (the **dependent variable**) **against** x (the **independent variable**)
- A straight-line graph **has an equation of the form** $y = mx + c$
- Curved graphs **can be transformed** into linear graphs by plotting the appropriate functions on the **independent** x-axis
- Errors **in the gradient and intercept can be obtained by drawing the shallowest and steepest gradients and comparing these with the gradient and intercept of the line of best fit**
- Areas **under graphs can provide a specific physical meaning**
- Parabolic, reciprocal, inverse square, exponential **and** sine **and** cosine **curves are also used to represent relationships between physical quantities**

QUICK TEST

1. On a Cartesian graph, which axis represents the dependent variable?

2. What is the gradient and the value of the intercept of a straight-line graph with equation $y = mx + c$?

3. What type of graph is represented by the following equations? (i) Ohm's Law $V = IR$ (a plot of V against I) (ii) Kinetic energy $E = \frac{1}{2}mv^2$ (a plot of E against v) (iii) Energy equation $E = \frac{hc}{\lambda}$ (a plot of E against λ)

4. For the above three equations (given in question **3**), what quantities would be plotted (on the y- and x-axes) in order to obtain a straight-line graph, assuming that R, m, h and c are constants?

5. In the above graphs, give the term that would represent the gradient

6. If the displacement s, given by $s = ut + \frac{1}{2}at^2$, is plotted with s along the y-axis and t along the x-axis, what would be the shape of the curve?

7. If Einstein's equation $E = mc^2$ is plotted with E as the dependent variable and m as the independent variable, what would the gradient be?

8. If the resistance R is plotted as a function of $\frac{1}{\text{cross-sectional area}}$, i.e. $\frac{1}{A}$, what would be the gradient and what are the units of the gradient?

9. What does the area under the graph represent for a graph of power against time?

10. A small metal ball bearing is released in a vertical tube of light oil and its displacement measured as a function of time. What would the gradient of the graph represent?

PRACTICE QUESTIONS

1. An experiment is carried out to determine the value of an unknown resistor. The table shows the results of the experiment.

Current, I (A)	0.3	0.5	0.7	0.9	1.1	1.3
Potential difference, V (V)	1.2	1.9	3.1	3.6	4.7	5.2

The uncertainty in the current reading was ± 0.1 A and in the potential difference reading was ± 0.2 V.

a) Plot a graph of V against I. For each point, draw an error bar to represent the uncertainties. Draw the line of best fit. **[4 marks]**

b) Determine the gradient of the line of best fit and give its units. **[2 marks]**

c) Draw the shallowest and steepest line that goes through these points and determine the gradients of these lines. **[2 marks]**

d) Express the gradient with the associated uncertainty based on these results **[2 marks]**

2. A parachutist jumped from an aeroplane and the first 7 seconds of free fall was recorded as shown in the table.

Time, t (s)	0	1	2	3	4	5	6	7
Velocity, v (m s^{-1})	0	4.6	6.9	7.6	7.8	8.0	8.0	8.0

a) Plot a graph of v against t between 0 and 7 seconds. **[4 marks]**

b) From the graph determine the gradient at (i) $t = 0.5$ s (ii) $t = 2$ s and (iii) $t = 6$ s. **[3 marks]**

c) What does the gradient represent? Give its units. **[2 marks]**

d) Determine the area under the curve for time between (i) 0 and 1 s and (ii) 0 and 7 s. **[3 marks]**

e) What does the area under the curve represent? **[1 mark]**

3. The kinetic energy of a car of mass 1 tonne is determined as a function of its speed on a straight track. The table shows the data that were obtained.

Speed, v (m s^{-1})	5	10	15	20	40	60	100
Kinetic energy, E (J)	12 000	47 000	115 000	195 000	710 000	1 855 000	4 875 000

a) Plot a graph of the kinetic energy (J) as a function of the speed, v, in m s^{-1}. **[4 marks]**

b) From the nature of the graph determine the equation that connects kinetic energy with velocity. **[2 marks]**

c) Re-plot the graph to obtain a straight-line relationship and determine the gradient of the line. **[3 marks]**

d) What does the gradient of the line represent? **[1 mark]**

Nuclei and Particle Decay

The Nucleus

At the centre of every atom there is a **nucleus** containing **nucleons**, i.e. **protons** and **neutrons**. Although atoms are of the order of 1×10^{-10} m (0.1 nm) in diameter, the nucleus is nearer 1×10^{-15} m or 0.000 001 nm in diameter (1 fm). However, almost all of the mass resides in the nucleus. The **mass of the proton and neutron** are nearly identical (1.67×10^{-27} kg), while the **electron mass** is 9.11×10^{-31} kg, i.e. some **2000** times smaller.

Nuclide notation is a method that summarises the information about the structure of an atom and is used in writing nuclear equations showing decay products. The **mass number** (A), also known as the **nucleon number**, is written at the top, and the **atomic number** (Z), also called the **proton number** is written at the bottom, next to the **element** symbol. $A = Z + N$, where N is the **neutron number**. In a neutral atom, there is an equal number of protons and electrons, with electrons characterising the chemical behaviour of the atom. Atoms that lose or gain electrons are called **ions** and an ion is symbolised with the inclusion of a $+$ or $-$ symbol next to the element, e.g. $^{23}_{11}Na^{+}$.

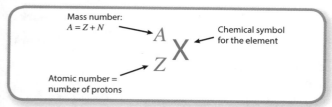

Mass number:
$A = Z + N$

Chemical symbol for the element

$^{A}_{Z}X$

Atomic number = number of protons

Isotopes

Atoms with the same number of protons but a different number of neutrons are called **isotopes**.

$^{1}_{1}H$ $^{2}_{1}H$ $^{3}_{1}H$

protium   deuterium   tritium

The number of neutrons dictates the **stability** of the nucleus; the larger or smaller the number of neutrons compared with the number of protons, the more unstable the nucleus becomes. (The chemical properties of a group of isotopes are unaffected by the number of neutrons within the nucleus.) Nuclei become more stable through the processes associated with **radioactive decay**.

A useful property of an element is called the **specific charge**, which is the ratio of its charge to its mass ($\frac{charge}{mass}$). Specific charge of a nucleus, a charged particle or an ion is defined as the ratio of its charge (Q) to its mass (m) i.e. Q/m. The units for specific charge are $C\,kg^{-1}$. The measurement of specific charge is one method by which nuclei, particles and ions can be identified. The specific charge of the neutron is 0.

Fundamental Forces within the Nucleus

There are only two significant fundamental forces that act on nucleons to hold them together within the nucleus. These two forces balance each other at a precise distance, allowing nucleons to remain stable inside the nucleus. They are:

- the **electrostatic (or electromagnetic) repulsive force** acting between protons that extends well beyond 10 fm
- the **strong nuclear force** acting between nucleons that is repulsive at very small distances (< 0.4 fm) but attractive between 0.4 fm and about 3 fm.

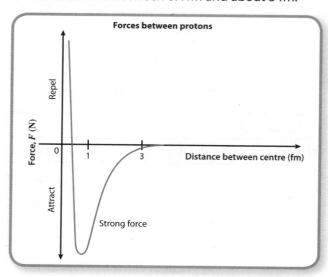

Forces between protons

A third attractive force, the **gravitational force**, is also present within the nucleus and binds nucleons together solely due to their mass. However, because nucleon masses are $\sim 10^{-27}$ kg, the gravitational force is extremely small. The fourth fundamental force, the **weak force**, plays a role in the decay of unstable nuclei.

Alpha Decay

Because the strong nuclear force has only a small range, nuclei with an excess of neutrons or protons are unstable and decay to restore stability. Heavy nuclei (those containing a large number of nucleons) are by their very nature unstable and stability is regained by emitting a large particle called an **alpha particle** (**nucleus of a helium atom**), which can be represented by $^4_2$He or $^4_2\alpha$. **Alpha decay** is the dominant process in nuclei containing more than **82 protons**; it reduces the nucleon number by 4 and the proton number by 2, with a change in element from X to Y:

$$^A_ZX \rightarrow {^{A-4}_{Z-2}}Y + {^4_2}\alpha$$

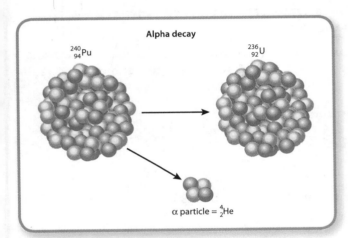

Alpha decay

$^{240}_{94}$Pu

$^{236}_{92}$U

α particle $= {^4_2}$He

An example would be the decay of plutonium-240 as shown above; also the decay of uranium-238 and the decay of radium-226 in granite rocks.

Beta Decay

For lighter nuclei, the dominant decay processes involve the ejection of smaller and lighter particles, in **beta-minus** or **beta-plus decay** via the so-called **weak interaction**. These processes allow neutron-rich nuclei to exchange their neutrons for protons and proton-rich nuclei to exchange their protons for neutrons. **Beta-minus decay** emits a beta-minus particle (which is identical to an atomic **electron** except that it is emitted from a nucleus) along with another fundamental particle called the **antineutrino** (strictly the electron antineutrino, $\overline{v}_e$). In this process, one of the neutrons is changed into a proton (the proton number increases by one but the nucleon number stays the same) with the antineutrino carrying away the **excess energy and momentum**. The nuclear equation for this decay is:

$$^A_ZX \rightarrow {^A_{Z+1}}W + {^0_{-1}}\beta + \overline{v}$$

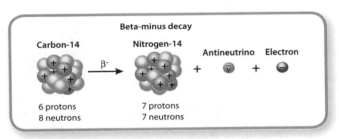

Beta-minus decay

Carbon-14 Nitrogen-14 Antineutrino Electron

β^-

6 protons 7 protons
8 neutrons 7 neutrons

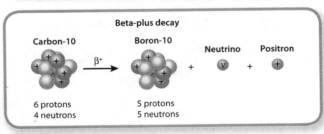

Beta-plus decay

Carbon-10 Boron-10 Neutrino Positron

β^+

6 protons 5 protons
4 neutrons 5 neutrons

An example would be the decay of carbon-14 (an isotope of carbon) to nitrogen-14. **Beta-plus decay** results in the emission of a **positive electron** (called a **positron**) and a **neutrino** (the electron neutrino, v_e). In this process, one of the protons is changed into a neutron (the proton number decreases by one but the nucleon number remains the same) with the emission of a neutrino. The nuclear equation is:

$$^A_ZX \rightarrow {^A_{Z-1}}V + {^0_1}\beta + v$$

An example would be the decay of carbon-10 (an isotope of carbon) to boron-10.

SUMMARY

SUMMARY

- **Nucleon** is the term used to collectively describe protons and neutrons; these particles form the nucleus

- The proton number or atomic number is the number of protons in the nucleus, and the mass number or nucleon number is the number of protons and neutrons in the nucleus

- Isotopes are different forms of the same element having the same number of protons but different number of neutrons

- Specific charge is the ratio $\frac{\text{charge } (Q)}{\text{mass } (m)}$ with units of $C\,kg^{-1}$

- The strong nuclear force is a very short-range force that acts between nucleons and holds the nucleus together; the other fundamental forces are electromagnetic, weak and gravitational forces

- The neutrino is a neutral and almost massless fundamental particle that rarely interacts with matter; antineutrinos are the antiparticles of the neutrino

- Unstable nuclei become more stable through the emission of alpha (helium nuclei) or beta (electrons emitted from the nucleus) particles and by gamma emission which removes excess energy

QUICK TEST

1. State the number of protons and neutrons in a nucleus of $^{24}_{11}\text{Na}$.

2. How many neutrons, protons and electrons are there in an atom of uranium-235?

3. What are isotopes?

4. The most stable isotope of copper is copper-63. Write this in the form of $^{A}_{Z}\text{X}$.

5. Copper also has the isotope copper-65. Write this in the form of $^{A}_{Z}\text{X}$.

6. Which particles in the nucleus have zero specific charge?

7. Calculate the specific charge of a beryllium-9 nucleus (contains five neutrons and four protons).

8. Calculate the specific charge of a lithium-7 ion (contains four neutrons, three protons and two electrons).

9. A $^{63}_{29}\text{Cu}$ atom loses two electrons and forms an ion; calculate the charge of the ion in coulombs and determine the specific charge of this ion.

10. Which force holds nucleons together in a nucleus?

11. What is the name of the force that tends to make a nucleus unstable?

12. Complete the following nuclear equation: $^{229}_{90}\text{Th} \rightarrow \quad \text{Ra} + \quad \alpha$.

13. Complete the following nuclear equation: $^{14}_{6}\text{C} \rightarrow \quad \text{N} + \quad \beta + \quad \bar{\nu}$.

PRACTICE QUESTIONS

1. a) Explain what is meant by the specific charge of a nucleus. [1 mark]

b) Calculate the specific charge of the nucleus of ^1_1H given that it has a charge of 1.60×10^{-19} C and a mass of 1.67×10^{-27} kg. [2 marks]

c) State which of the particles proton, neutron and electron has:

(i) no specific charge [1 mark]

(ii) the smallest specific charge. [1 mark]

d) An atom of magnesium $^{24}_{11}Mg$ is ionised by removing an electron.

(i) State the number of protons, neutrons and electrons in the ion formed. [2 marks]

(ii) Calculate the charge on the ion. [1 mark]

(iii) Calculate the specific charge of the ion. [2 marks]

2. The equation $^{65}_{28}Ni \rightarrow {}^A_Z Cu + {}^{\ 0}_{-1}\beta + X$ represents the decay of nickel-65 by the emission of a β^- particle.

a) Identify the particle X. [1 mark]

b) Determine the values of A and Z. [2 marks]

c) What is another name given to a β^- particle? [1 mark]

d) What does the emission of a β^- particle indicate about the nucleus of nickel? [1 mark]

e) The β^- particle is emitted with a range of energies. What does this indicate about the unknown particle X and the total energy emitted? [2 marks]

3. a) Describe how the strong nuclear force between two protons varies with the distance of separation of the two protons, giving suitable values for the separation. You may find a sketch useful. [5 marks]

b) Explain why very heavy nuclei that have too many protons decay, and state and explain what mechanisms take place to restore stability. [3 marks]

c) Copy and complete the nuclear equation that shows the decay of uranium-238 by the above decay process. $^{238}_{92}U \rightarrow \quad Th + \quad He$ [2 marks]

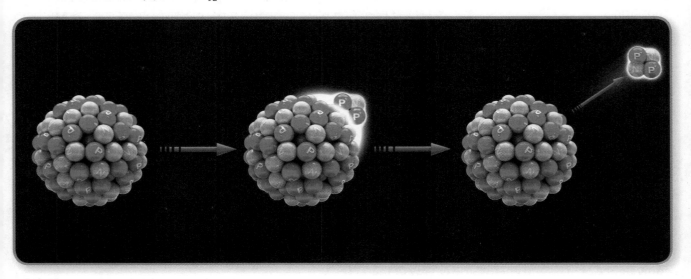

Particles, Antiparticles, Exchange Particles and Quarks

Particles and Antiparticles

Matter is made of **atoms** and atoms are made of **fundamental particles**. For every **particle** there is a corresponding **antiparticle** that has the **same mass** but **opposite charge**. Antiparticles were predicted by Dirac (he predicted the existence of the **positron**, the antiparticle of the electron). The antiparticle of a proton is the **antiproton**, which has exactly the same mass but a negative charge. The neutrino and **antineutrino** are encountered in beta decay, and the neutron has an antiparticle called the **antineutron** with the same mass and equally neutral.

Particle	Symbol	Charge
proton	p	+1
neutron	n	0
electron	e^-	−1
electron neutrino	ν_e	0

Antiparticle	Symbol	Charge
antiproton	$\bar{p}$	−1
antineutron	$\bar{n}$	0
positron	e^+	+1
electron antineutrino	$\bar{\nu}_e$	0

Particle Annihilation

Today only particles remain. Antiparticles that were around after the **Big Bang** have all since disappeared. Antiparticles therefore have to be created in particle accelerators or are created in cosmic rays bursts that enter the upper atmosphere. They only live for a very short time (fractions of a second) before an antiparticle meets a particle resulting in the **annihilation** of both particles. The mass of both particles is turned into energy according to Einstein's famous equation $E = mc^2$, which states the **equivalence** of mass and energy. The energy of the annihilation produces **two gamma-ray photons** that move off in opposite directions to each other (an effect utilised in PET scanners). The minimum energy of the photons produced by annihilation is the **rest energy** of the two particles involved, i.e. $2E_0 = 2m_oc^2$; the rest energy of an electron (or positron) is 0.511 MeV.

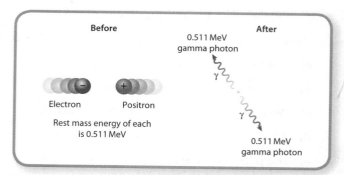

Before / After

0.511 MeV gamma photon

γ

Electron   Positron

Rest mass energy of each is 0.511 MeV

γ

0.511 MeV gamma photon

Pair Production

The opposite of annihilation can also happen. When energy in the form of a gamma-ray photon is converted into mass, two particles are created, a **particle–antiparticle pair**, in a process called **pair production**. Pair production can only happen if the gamma-ray photon has sufficient energy to produce the total mass of both particles. It also has to take place close to a more massive object such as a nucleus in order to conserve momentum (the heavy nucleus recoils a little). The minimum energy needed for pair production is therefore the total rest energy of the particles that are produced:

$$E_{min} = 2E_0$$

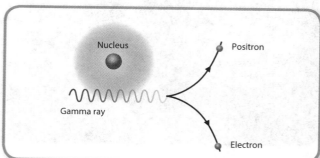

Nucleus   Positron

Gamma ray

Electron

Quarks

Quarks are another type of fundamental particle in addition to the electron, positron, neutrino and antineutrino. They are the building blocks that make protons and neutrons and as such are subject to the strong nuclear force. There are six kinds of quarks, each with their corresponding antiparticle (antiquark). The three most important quarks (and their antiquarks) are **up** (u), **down** (d) and **strange** (s), and **anti-up** $(\bar{u})$, **anti-down** $(\bar{d})$ and **anti-strange** $(\bar{s})$. Particles made from quarks are collectively known as **hadrons**.

Baryons

Protons and neutrons are made using **three quarks** (qqq). We call such particles **baryons** and these particles are subject to the strong nuclear force. Particles made from **three antiquarks** $(\overline{qqq})$ are called **antibaryons**. The basic properties of quarks and antiquarks are shown in the table. **Protons** are made from (uud) quarks, giving them a total **charge** of $+\frac{2}{3} + \frac{2}{3} - \frac{1}{3} = +1$. The antiproton is therefore made from $(\overline{uud})$ quarks, giving them a total charge of -1. A **neutron** has the quark structure (udd) and the **antineutron** $(\overline{udd})$, both having a charge of 0. Other baryons can be made using not just the up and down quarks but also the strange quark and anti-strange quark, as long as three quarks (or antiquarks) are used. This gives rise to a host of other more exotic particles. For example, the sigma particle Σ^+ is made from (uus) and Σ^0 is made from (uds).

	Name	Symbol	Charge	Mass (MeV c^{-2})
Quarks	up	u	$+\frac{2}{3}$	1.5–3.3
	down	d	$-\frac{1}{3}$	3.5–6.0
	strange	s	$-\frac{1}{3}$	70–130
Antiquarks	anti-up	$\bar{u}$	$-\frac{2}{3}$	1.5–3.3
	anti-down	$\bar{d}$	$+\frac{1}{3}$	3.5–6.0
	anti-strange	$\bar{s}$	$+\frac{1}{3}$	70–130

Mesons and the Strange Quark

Particles can also be made from just two quarks in a special **quark–antiquark pairing** $(q\bar{q})$; these are collectively called **mesons**. Two types of mesons exist: those that do not involve the strange quark are called **pions** and those that do are called **kaons**. There are three types of **pions** with the following quark structures: $\pi^0 (u\bar{u})$ or $(d\bar{d})$, $\pi^- (d\bar{u})$ and $\pi^+ (u\bar{d})$. There are four types of **kaons**: $K^0 (d\bar{s})$, $\overline{K}^0 (s\bar{d})$, $K^+ (u\bar{s})$ and $K^- (s\bar{u})$.

The reason the neutral kaon has two distinct forms is that they can be constructed from two different quark arrangements that satisfy the quark structure of kaons but are also both neutral. To distinguish between them, the kaon containing the down quark is a called a neutral kaon and the one containing the anti-down quark is known as an anti-neutral kaon.

Quark Confinement

Evidence for the quark structure of particles comes from **high-energy electron interactions** with protons. Despite these energetic collisions, no free quarks have been observed, only quark–antiquark pairs are created; this is known as **quark confinement**. Quarks explain the process of beta-plus and beta-minus decay; for example, in **beta-minus decay**, a neutron (udd) is changed into a proton (uud). In terms of quarks, a down quark is changed into an up quark and any change in quarks is governed by the **weak interaction**.

Exchange Particles

Finally, the fundamental forces of nature are mediated by another type of particle called **gauge bosons**. The mass of particles is created by the Higgs field, mediated by the **Higgs boson**. These exchange particles and their properties, according to the Standard Model of particle physics, are listed in the table; note that the exchange particle for quarks are very short range **gluons**.

Exchange particle	Symbol	Antiparticle	Charge	Mass (MeV c^{-2})	Interaction
photon	γ	self	0	0	electromagnetic
W boson	W^+	W^-	-1	80	weak
Z boson	Z_0	self	0	91	weak
gluon	g	self	0	0	strong
Higgs boson	H^0	self	0	125	mass
graviton	G	self	0	0	gravitation

SUMMARY

- An antiparticle is identical in all aspects to its particle except that it has opposite charge
- Particles and antiparticles have rest energy in MeV, rest mass in MeV c^{-2} and charge in C
- When a particle meets its antiparticle, they annihilate each other; their total mass is converted into energy in the form of two gamma-ray photons
- Pair production is the creation of a particle–antiparticle pair in the presence of a nucleus from a high-energy photon provided the energy is large enough
- The minimum energy for pair production is the combined rest energies for the electron and positron, i.e. $2 \times 0.511\,\text{MeV} = 1.022\,\text{MeV}$
- Quarks are fundamental particles that make up protons and neutrons; they exert the strong nuclear force on each other
- There are six types of quarks in total (and their six corresponding antiquarks) that make up composite matter
- Only the quark flavours up, down, strange and anti-up, anti-down and anti-strange are required in A-level courses
- Quarks have fractional charges of $+\frac{2}{3}$ (u), $-\frac{2}{3}$ ($\overline{\text{u}}$), $-\frac{1}{3}$ (d), $+\frac{1}{3}$ ($\overline{\text{d}}$), $-\frac{1}{3}$ (s) and $+\frac{1}{3}$ ($\overline{\text{s}}$)
- Exchange particles are involved in interactions between particles; they are only created, emitted and absorbed between interacting particles and are the mediators of the fundamental forces
- Exchange particles are also called gauge bosons or virtual bosons

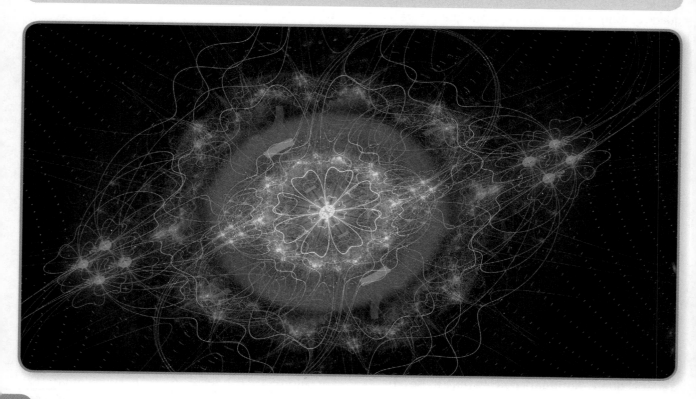

1. What is the minimum energy required to produce two photons from the annihilation of an electron–positron pair?
2. If a high-energy photon has an energy greater than 1.022 MeV and produces an electron–positron pair, what happens to the excess energy?
3. What is the minimum energy produced in a proton–antiproton annihilation (the rest energy of a proton is 938 MeV)?
4. What force governs the behaviour of baryons?
5. What are baryons made from?
6. What do we call particles made from quark–antiquark pairs?
7. Name two types of mesons.
8. Which particles are made from strange quarks?
9. Which are the heavier particles, pions or kaons?
10. How many types of pions are there, and what are they?
11. What are the quark structures of a neutron and of a K^+ kaon?
12. Why are there two types of K^0 particle?
13. What particles are created and destroyed in the interactions between particles?

PRACTICE QUESTIONS

1. Protons and neutrons are part of a group called the hadrons that are composed of quarks. There are two sub-groups of hadrons known as baryons and mesons.
   a) What is the fundamental property that defines a hadron? **[1 mark]**
   b) State the quark structure of a baryon and give an example. **[2 marks]**
   c) State the quark structure of a meson. **[1 mark]**
   d) Mesons themselves form two further sub-groups. Name these sub-groups and explain the difference between them. **[3 marks]**

2. a) What is the name given to the class of particles represented by the combination of three antiquarks ($\overline{qqq}$)? **[1 mark]**
   b) Name an example of a particle made from three antiquarks and give its quark structure. **[2 marks]**
   c) For this particle, state its charge and show how its charge is determined from its quark structure. **[2 marks]**
   d) Explain why the antiparticle is not found in ordinary matter. **[2 marks]**

3. a) (i) Define what is meant by a quark. **[2 marks]**
      (ii) Explain the term 'quark confinement'. **[1 mark]**
   b) Name the exchange particle that acts between quarks and comment on the distance over which it interacts. **[2 marks]**
   c) Arrange the following particles in order of their masses (lightest first): proton, electron, up quark, neutron, down quark. **[1 mark]**
   d) The antineutron is the antiparticle of the neutron. State one difference and one similarity between a neutron and antineutron. **[2 marks]**
   e) For the antineutron give: (i) its quark structure; (ii) its charge. **[2 marks]**

The Standard Model and Conservation Rules

The Standard Model

The **Standard Model** is used to describe the nature of matter that is composed of particles, both **fundamental** and **composite**. **Cosmic rays** from outer space and **particle accelerators** provide a means of creating particles and antiparticles that are rarely observed, and their decay products. There are four fundamental forces of nature that allow matter, and hence particles, to interact; they are are the **electromagnetic**, **strong**, **weak** and **gravitational force**. These forces are transmitted via **exchange particles** (also called **virtual bosons** or **gauge bosons**) that consist of the **photon**, **gluon**, **W/Z** and the **graviton**, respectively. The **exchange particles**, **leptons** and **quarks** are the fundamental particles that are in the Standard Model; all other particles are composite particles made from these fundamental particles.

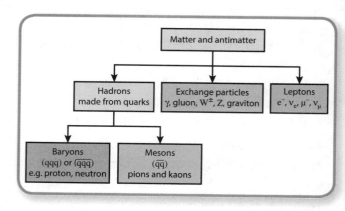

Fundamental Forces

Force	Particles experiencing force	Force carrier particle	Range	Relative strength
Gravity acts between objects with mass	all particles with mass	graviton (not yet observed)	infinite	much weaker
Weak force governs particle decay	quarks and leptons	W^+, W^-, Z^0 (W and Z)	very short range	
Electromagnetism acts between electrically charged particles	electrically charged	γ (photon)	infinite	
Strong force binds quarks together	quarks and gluons	g (gluon)	short range	much stronger

Leptons and Lepton Number

Leptons comprise the **electron** (e⁻), which is a stable particle, and its more massive counterpart, the **muon** (μ^-), which is unstable and decays eventually into an electron, and **neutrinos**. Both electrons and muons have associated neutrinos, ν_e and ν_μ. Neutrinos have almost zero mass and have zero charge and therefore interact only extremely weakly with matter through the weak interaction. Leptons do not 'feel' the strong nuclear force and are only subject to the **weak interaction** (and, under certain circumstances, the electromagnetic and gravitational force as well). These four fundamental leptons also have their antiparticle equivalents, the **positron** (e⁺), the **antimuon** (μ^+) and the two **antineutrinos** $\bar{\nu}_e$ and $\bar{\nu}_\mu$. There is a third tier in the Standard Model where the **tau particle** (and associated neutrino) exist. However, they are not part of exam specifications and are only mentioned for completeness.

A particle is assigned a **lepton number**, L, of +1, −1 or 0 depending on whether it is a lepton, an antilepton or a non-lepton. Electrons, muons and the electron and muon neutrinos are assigned $L = +1$; positrons, antimuons and the electron and muon antineutrinos are assigned $L = -1$; all other particles (i.e. non-leptons) are assigned $L = 0$. In any interaction, not only is total lepton number conserved but also the individual electron and muon lepton numbers.

Hadrons

Particles called **hadrons** are not fundamental particles because they are made up of **quarks**; as such, they feel the strong nuclear force. There are two types of hadrons: **baryons** and **mesons**. The most familiar members of the baryon family are the **proton** and **neutron**. All baryons are unstable except for the proton and eventually decay into a proton and other particles. The antiparticles of protons and neutrons (antiprotons and antineutrons) are antibaryons.

In a similar way to leptons, baryons are assigned a number called the **baryon number**, B. The proton and neutron (and all other baryons) are assigned $B = +1$; antibaryons are assigned $B = -1$; all other particles (non-baryons, such as mesons, i.e. pions and kaons) are given the baryon number $B = 0$. In the same way as the lepton number is conserved in an interaction, the total baryon number in any particle interaction is also conserved. The above rules also mean that a baryon number of $+\frac{1}{3}$ can be assigned to any quark, $-\frac{1}{3}$ to any antiquark and 0 to any lepton.

Conservation of Baryon and Lepton Number

In all particle interactions, both **energy** and **momentum** are conserved. In addition, **charge** is conserved, as are both the **baryon number** and the **lepton number**. If these conservation rules are not obeyed then the particle interaction cannot take place. The application of these rules is just a matter of checking both sides of the nuclear equation to see whether they balance. Take the following decay mode of a kaon as an example:

$$K^- \rightarrow \mu^- + \bar{\nu}_\mu$$

● Charge (Q): $-1 = -1 + 0$

● Baryon number (B): $0 = 0 + 0$

● Lepton number (L): $0 = +1 - 1$

As all conservation rules apply, this interaction is possible.

Conservation of Strangeness

Strangeness is only conserved for strong interactions involving particles that contain a **strange quark** (and/or anti-strange quark), such as kaons. A **strangeness number** of $S = -1$ is assigned to the strange quark and $S = +1$ for the anti-strange quark. This means that the K^- meson has $S = -1$, as does the $\bar{K}^0$ meson. Non-strange particles (e.g. protons, neutrons, pions, leptons…) are assigned the value $S = 0$. However, if the weak interaction is involved then strangeness is not conserved, as in the K^- meson decay in the example below left.

	Quarks			Antiquarks		
	up	down	strange	anti-up	anti-down	anti-strange
	u	d	s	$\bar{u}$	$\bar{d}$	$\bar{s}$
strangeness, S	0	0	−1	0	0	+1

SUMMARY

- All the leptons, quarks and exchange particles make up the particles in the Standard Model
- The four fundamental interactions are: strong, weak, electromagnetic and gravitational
- Baryon number, B, is always conserved in particle interactions; the baryon number assigned to a quark is $+\frac{1}{3}$ and to an antiquark is $-\frac{1}{3}$
- The proton is the only stable baryon; all other baryons decay into protons
- Free neutrons are unstable and decay via the weak interaction into protons
- Leptons are subject to the weak interaction but not to the strong interaction
- Lepton number, L, is always conserved in particle interactions
- Muons decay into electrons and positrons
- Strange particles are produced through the strong interaction and are always created in pairs to conserve strangeness
- Particles containing strange quarks decay through the weak interaction
- Strangeness is conserved in strong interactions but not if the weak interaction is involved
- Conservation rules for energy, momentum, charge, baryon number, lepton number and strangeness can be applied to all particle interactions

STANDARD MODEL OF ELEMENTARY PARTICLES

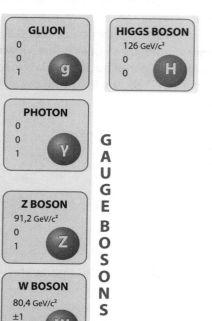

QUARKS

	UP	DOWN	STRANGE
mass	2,3 MeV/c²	4,8 MeV/c²	95 MeV/c²
charge	⅔	-⅓	-⅓
spin	½ u	½ d	½ s

GLUON
0
0
1 g

HIGGS BOSON
126 GeV/c²
0
0 H

PHOTON
0
0
1 γ

LEPTONS

ELECTRON	MUON	ELECTRON NEUTRINO
0,511 MeV/c²	105,7 MeV/c²	<2,2 eV/c²
-1	-1	0
½ e	½ μ	½ νₑ

MUON NEUTRINO	TAU NEUTRINO
<0,17 MeV/c²	<15,5 MeV/c²
0	0
½ νμ	½ ντ

Z BOSON
91,2 GeV/c²
0
1 Z

W BOSON
80,4 GeV/c²
±1
1 W

GAUGE BOSONS

1. The Standard Model is based on which types of fundamental particles?

2. To which group of particles do the following particles belong: (i) neutron (ii) muon (iii) K^0?

3. State the baryon number of: (i) muon (ii) antiproton (iii) up-quark.

4. State the lepton number of: (i) neutron (ii) neutrino (iii) antimuon.

5. Is the following particle interaction possible? $p + \bar{v}_e \rightarrow e^+ + n$

6. Apply the conservation of lepton number to the following decay process and state whether the decay process can occur: $n \rightarrow p + e^- + \bar{v}_e$

7. The omega minus particle (Ω^-) is a baryon with a strangeness of −3 and a charge of −1. Determine its quark structure.

8. A K^- meson was found to decay into three charged π mesons. (i) What type of interaction takes place in this interaction, and (ii) is strangeness conserved?

9. The following interaction is observed: $p + p \rightarrow p + n + \pi^+$; (i) show that charge is conserved and (ii) give another conservation rule that is also obeyed.

10. Why is the following interaction not possible? $p + \bar{p} \rightarrow p + \pi^-$

PRACTICE QUESTIONS

1. The neutron is an example of a hadron and an electron neutrino is an example of a lepton.

    a) State one similarity and one difference between a neutron and an electron neutrino. **[2 marks]**

    b) There are four fundamental forces of nature. Name these four forces and give the exchange particle that mediates each force. **[4 marks]**

    c) Which of the four forces is involved in the interaction between a neutron and an electron neutrino? **[1 mark]**

    d) Write the interaction equation between a neutron and an electron neutrino, ensuring that the conservation rules apply. **[2 marks]**

2. The proton belongs to the hadron family of particles.

    a) Describe the key properties of a hadron. **[2 marks]**

    b) Hadrons also form two other sub-groups of particles. Name these two groups and describe their composition in terms of quarks. **[2 marks]**

    c) Give an example of a particle in each of these sub-groups, along with its quark structure. **[2 marks]**

3. The lambda zero particle, Λ^0, is a baryon with zero charge and strangeness of −1.

    a) Write down the possible quark composition of Λ^0. **[1 mark]**

    b) Lambda zero may decay via the following process: $\Lambda^0 \rightarrow p + \pi^-$. Using the conservation rules, verify that such a decay process does occur. **[4 marks]**

    c) Why is the conservation of strangeness not obeyed? **[1 mark]**

    The decay of lambda zero can also produce a neutron in the decay mode $\Lambda^0 \rightarrow n + X$.

    d) Identify particle X and write down its quark structure. **[2 marks]**

    e) State one difference and one similarity between particle X and π^-. **[2 marks]**

Particle Interactions and Feynman Diagrams

Exchange Particles

When particles decay or interact, they do so by exchanging a particle that carries the force of the interaction between the particles. This energy exchange manifests itself as an **exchange particle or boson** (also called a **virtual particle**) that lives only for a very short time (long enough to exchange energy, momentum, and other properties). The range of the force is inversely proportional to the mass of the exchange particle. The **W boson** involved in beta decay is 100 times more massive than a proton and so only has a range of ~ 0.001 fm. The exchange particles involved between quarks are called **gluons** and those between charged particles are called **photons**. All of these have been directly or indirectly observed except for the exchange particle involving gravitation, the **graviton**, which has so far evaded detection.

Feynman Diagrams

Feynman diagrams are a way to 'picture' an interaction or decay and they were introduced by Richard Feynman to simplify the interpretation of interactions. The basic rules for constructing Feynman diagrams are: incoming particles move from the bottom upwards (along an imaginary time axis) with baryons on the left side of the diagram and leptons on the right side; such particles are drawn as straight lines; exchange particles are shown as wavy lines except for gluons that are often shown as spirals; because of their

short lifetime exchange particles may be shown either horizontally or at an upward angle; arrows are used to represent the direction of the particles.

For example, **beta-minus decay** (a weak interaction) involves a neutron transforming into a proton, i.e. ${}^{1}_{0}n \rightarrow {}^{1}_{1}p + {}^{0}_{-1}e + \bar{\nu}_e$. The Feynman diagram for this interaction reveals more detail than that given by the nuclear equations. Interactions involving particles made from quarks may have individual quark lines drawn parallel and next to each other; during beta-minus decay, a down quark changes into an up quark, as shown in the diagram. Because the lifetimes of the exchange particles are generally very short (of the order of 10^{-25} s), they are shown on Feynman diagrams as horizontal or angled wavy lines.

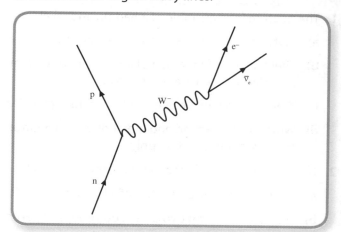

Fundamental force	Exchange boson	Mass (in MeV c^{-2})	Expected range
Electromagnetic	photon, γ	$0\ (< 3 \times 10^{-33})$	infinite
Weak	W^+, W^-, Z^0	$W^{\pm} = 80 \times 10^3$ $Z^0 = 91 \times 10^3$	very short range, ~ 0.001 fm
Strong	gluon, g	not observable as an individual particle, expected mass $= 0$	short range, ~ 1 fm
Gravity	graviton, G	not proven to exist, expected mass $= 0$	infinite

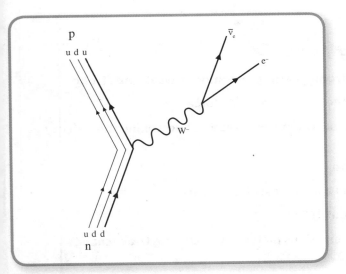

Particle Lifetimes

The **lifetime** of a particle is the mean time between being created and being annihilated. These are very short for the exchange particles but they are much longer for **pions** (10^{-17} s to 10^{-8} s) and longer still for the **muon** (2.2 μs). A **free neutron** decays into a proton after ~620 s whereas both the **electron** (10^{26} s) and **proton** (10^{29} s) are stable (as far as we know).

Other Particle Interactions

A number of other important interactions can also be depicted in Feynman diagram form.

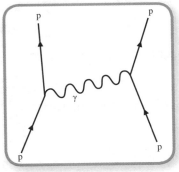

A Feynman diagram of **proton–proton interactions** shows the exchange of a **virtual photon** (indicated by the photon symbol γ) to represent the **electromagnetic** interaction.

Electron–electron interactions are another example of how, when two particles of equal charge interact, they repel each other through the exchange of a **virtual photon** in an **electromagnetic** interaction.

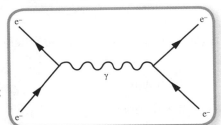

Electron capture is an interaction that is the opposite of beta-minus decay and an example of the **weak** interaction. In this case, a proton-rich nucleus 'captures' or absorbs an electron from an atom. The proton–electron interaction produces a neutron and emits a neutrino to conserve lepton number. The exchange particle in this case involves the W^+ boson with an arrow to indicate that the proton acts on the electron.

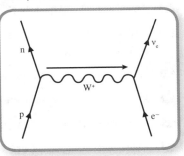

Electron–proton collisions occur when high-energy electrons interact with protons:

$$^1_1p + {}^0_{-1}e \rightarrow {}^1_0n + \nu_e$$

In this case, it is the electron (negative charge) that interacts with the proton via the **weak** interaction, as indicated by the exchange particle W^-; as the electron collides with the proton, an arrow is included to indicate the direction of movement of the W^- particle.

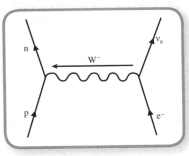

Binding of nucleons: the **gluon** is the exchange particle that binds nucleons together in the nucleus via the **strong** interaction. The Feynman diagram for proton–neutron binding is shown with the exchange particle indicated by a spiral curve. You will not be required to recall information specifically about either the gluon, the graviton or the exchange particle Z^0 that is involved in some weak interactions.

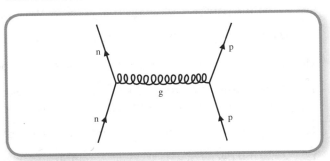

SUMMARY

- The four fundamental forces of nature are: electromagnetic, gravitational, weak and strong
- The virtual photon is the exchange particle for the electromagnetic force
- Particle lifetime is the mean time that a particle exists between creation and annihilation, decay or interaction
- Fundamental particles are stable, as is the proton
- A free neutron has a lifetime of $\sim$15 minutes and a muon lasts for $\sim$2.2 μs
- Lifetimes of exchange particles are typically about 10^{-25} s
- Feynman diagrams are used to show particle interactions pictorially, including the exchange particles
- The W^+ and W^- are the exchange particles involved in the weak interaction; they are shown by a wavy line in a Feynman diagram
- Gluons are the exchange particle for the strong nuclear force or strong interaction and these are shown as a continuous spiral in a Feynman diagram
- Electron capture, electron–proton collisions and beta-minus and beta-plus decay are examples of the weak interaction

QUICK TEST

1. What is a Feynman diagram?

2. How are matter particles represented on a Feynman diagram?

3. How are exchange particles shown?

4. How are particles that are created and annihilated shown on a Feynman diagram?

5. Why are exchange particles known as virtual particles?

6. Draw a Feynman diagram to illustrate β^- decay, i.e. $n \rightarrow p + e^- + \bar{v}_e$

7. What is the exchange particle for beta decay?

8. Exchange particles generally transfer from which direction?

9. What is a gluon and how is this shown on a Feynman diagram?

10. What interaction is involved in all muon decay processes?

11. Muons (heavy electrons) always decay into an electron (or positron) and which two other particles? (hint: consider electron and muon lepton numbers)

12. Which exchange particle represents the electromagnetic interaction?

13. Show the decay of a μ^- as: (i) an equation and (ii) as a Feynman diagram.

PRACTICE QUESTIONS

1. Electron scattering occurs when two electrons interact with each other.

   a) Draw and label a Feynman diagram to illustrate the interaction between electrons. **[3 marks]**

   b) Give the name of the force mediating this interaction and the name of the exchange particle involved. **[2 marks]**

   c) As in all interactions, both energy and momentum are conserved. Give the name of another three quantities that are conserved in the above interaction, indicating the conservation equation. **[3 marks]**

2. β^+ decay occurs in materials possessing too many protons.

   a) Write a nuclear equation that describes this decay process. **[2 marks]**

   b) Draw a Feynman diagram for β^+ decay, labelling the five particles involved. **[3 marks]**

   c) Give the name of the fundamental interaction involved in β^+ decay and the name of the exchange particle that mediates this force. **[2 marks]**

   d) Explain in terms of the quark structure what happens in the decay process. **[1 mark]**

   e) Using three conservation laws (other than energy and momentum), show that the decay process given in part (a) can occur. **[3 marks]**

3. An electron may collide with a proton as shown in the partly completed Feynman diagram below.

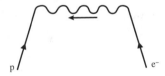

   a) Complete the Feynman diagram and label the missing particles. **[3 marks]**

   b) Name the fundamental force responsible for this interaction **[1 mark]**

   c) Write a nuclear equation to describe this interaction **[2 marks]**

   d) Name three other properties that are conserved in this interaction apart from energy and momentum and use these laws to show that the interaction can occur. **[3 marks]**

Photons and the Quantum of Light

The Photoelectric Effect

The **photoelectric effect** is the emission of 'free' electrons near the surface of a metal by the absorption of visible or ultraviolet radiation that break the bonds holding the electrons within the metal; it was first observed in the late 1880s. The observation created a major controversy in the physics world that contributed to the birth and development of quantum physics. The electrons emitted from the surface of a metal are called **photoelectrons** and the three key conclusions that result from photoelectric experiments are:

- for a given metal, **no photoelectrons** are emitted if the incident radiation has a frequency below the so-called **threshold frequency, f_0**

- photoelectrons are emitted with **kinetic energy, E_k,** ranging from zero (no emission) up to a maximum value that depends on the particular metal and on the **frequency** (energy) of the incident radiation but not on its intensity

- the **number of photoelectrons** emitted increases as the **intensity** of the radiation increases (provided the frequency of incident radiation is above the threshold frequency).

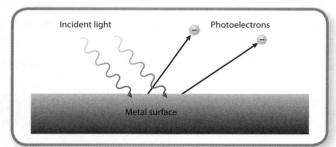

Incident light   Photoelectrons

Metal surface

Explanation Based on Light Quanta

The wave theory of light, based on wavefronts, was unable to explain the need for a threshold frequency or why there was a range of kinetic energies that depended only on the frequency and not on the intensity of the light.

Einstein proposed the **photon model of light** to explain the photoelectric effect. In this model, Einstein used the idea put forward earlier by Planck in which light and other types of electromagnetic radiation is emitted in small packets of energy called **quanta**, where the energy E is related to the frequency f through the equation $E = hf$; the constant of proportionality h is known as **Planck's constant**. The energy can also be expressed in terms of the wavelength of radiation, λ, using the wave equation $c = f\lambda$, which connects frequency with wavelength using the velocity of light, c. The word **photon** was only used much later to represent a quantum, or discrete amount, of light energy.

Einstein's model succeeded in providing an explanation for the key effects observed.

- The incident photon gives all of its energy to a single electron.

- If the incident energy is not large enough then the electron cannot escape from the surface of the metal and hence no photoelectrons are seen. It is only when the incident energy, $E = hf$ is greater than a particular **threshold energy** or corresponding **threshold frequency** that the effect is observed.

- If the **light intensity** increases, the number of photons also increases but the energy available to each electron remains the same. More incident photons interact with more electrons and hence more photoelectrons are released.

- The **threshold energy** (which varies for different metals) describes the minimum energy for an electron at the surface to be removed from a metal and is also called the **work function**, Φ.

- The work function or threshold energy can be modelled by a **potential well**, where the work function is equal to the minimum energy needed to escape the potential well.

Threshold Energy and Work Function

Each photon interacts with just one surface electron, giving up its entire energy in the interaction. Electrons in a metal can be described as being trapped inside a potential well. Before an electron can escape from the surface, it needs enough energy to break the bonds holding it to the atom and this finite amount of energy is called the **work function**, Φ; it is comparable to the **ionisation energy**.

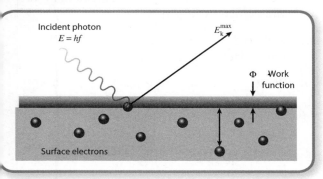

It is the least energy needed to release an electron from a metal and is related to the threshold frequency (or threshold energy) by the equation $\Phi = hf_0$. For electrons to be emitted, an energy greater than the work function is needed, i.e. $hf > \Phi$, and the minimum frequency (the threshold frequency) must be $f_0 = \dfrac{\Phi}{h}$.

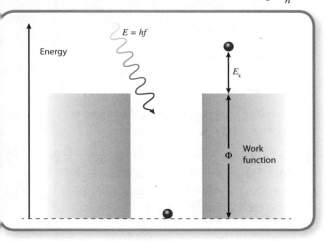

Kinetic Energy and Einstein's Photoelectric Equation

If the energy of the incident radiation is substantially greater than the threshold energy then not only will the electron escape from the surface but any excess energy will be shown as the **kinetic energy**, E_k, of the emitted photoelectron.

An electron directly on the surface of a metal will therefore be emitted with the **maximum kinetic energy** possible. The more deeply embedded surface electrons will need additional energy to escape and will therefore leave the surface with less than the maximum kinetic energy. This results in Einstein's photoelectric equation:

$$hf = \Phi + E_k^{\text{max}} \quad \text{or} \quad hf = \Phi + \frac{1}{2}mv_{\text{max}}^2$$

A graph of kinetic energy versus frequency will therefore give a straight line with a slope equal to Planck's constant and an intercept equal to the work function.

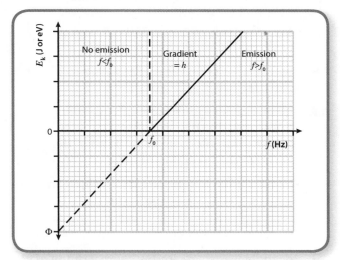

Potassium, which has a work function (threshold energy) of 2.3 eV (work functions are usually expressed in units of eV where $1\,\text{eV} = 1.6 \times 10^{-19}$ J), is a good example. Incident radiation of varying photon energies gives rise to either no emission of photoelectrons or emission with varying amounts of maximum kinetic energy.

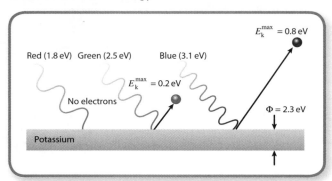

SUMMARY

- A quantum of light (a photon) is a packet of energy given by $E = hf$, where h is Planck's constant and f is the frequency; frequency and wavelength are related through the wave equation $c = f\lambda$

- The threshold frequency, f_0, is the minimum frequency required to cause the photoelectric effect and the emission of photoelectrons

- The threshold energy or work function, Φ, is the least energy needed to release a photoelectron from a metal; it is related to the threshold frequency by $\Phi = hf_0$

- The work function can be modelled as a potential well

- The photoelectric effect involves a one photon and one electron interaction in which a photoelectron absorbs all of the energy of the incoming photon

- The maximum kinetic energy an electron carries when it leaves the surface of a metal is given by $hf = \Phi + \frac{1}{2}mv^2_{max}$; this is Einstein's photoelectric equation

- Photoelectrons are emitted with a range of kinetic energies depending on how deeply embedded they are within the metal surface; electrons close to the surface attain the maximum kinetic energy

- The number of photoelectrons emitted is proportional to the intensity of the incident radiation

QUICK TEST

1. What name is given to the electrons that are emitted in the photoelectric effect?

2. What is meant by the photoelectric effect?

3. What is meant by the threshold frequency?

4. How is the threshold frequency related to the work function for a metal?

5. What is the equation that describes the photoelectric effect?

6. What is a quantum of light energy called?

7. What is the equation for maximum kinetic energy?

8. How do you convert from electronvolts to joules?

9. What are the two ranges of frequencies that are used to trigger the photoelectric effect in metals?

10. Why is the work function different for different metals?

11. Calculate the energy of a photon, in joules, that has a wavelength of 650 nm.

12. Sodium has a work function of 2.3 eV. Determine the threshold frequency for sodium metal.

13. What model is used to help understand the notion of the work function?

14. If blue light of energy 3.1 eV is incident on potassium metal (work function 2.3 eV), what is the maximum kinetic energy of the emitted electrons?

PRACTICE QUESTIONS

1. The threshold energy of light needed to cause the photoelectric effect in calcium is 4.64×10^{-19} J. Initially light with a wavelength of 0.4 µm is shone onto the calcium metal.

   a) Calculate the frequency (in Hz) of the incident radiation and its photon energy in eV. **[3 marks]**

   b) Determine the work function for calcium in eV. **[1 mark]**

   c) Calculate the maximum kinetic energy of the emitted photoelectrons. **[2 marks]**

   The light source is now changed so that the wavelength of the incident radiation is lowered to 0.35 µm.

   d) What effect would this have on the kinetic energy of the emitted photoelectrons? **[2 marks]**

2. A metal surface emits photoelectrons when a certain frequency of light is shone onto it.

   a) Explain why below a certain frequency no photoelectrons are emitted. **[3 marks]**

   b) Explain how this particular restriction on the frequency provides further supporting evidence against the wave theory of light to describe the photoelectric effect. **[2 marks]**

   c) The work function of iron is 7.2×10^{-19} J. Calculate the frequency above which photoelectrons will be emitted. **[2 marks]**

   d) A beam of photons is shone onto an iron sheet that emits photoelectrons with a maximum kinetic energy of 3.4×10^{-19} J. Determine the energy (in eV) and the frequency of the incident radiation. **[3 marks]**

3. The figure shows a graph of the kinetic energy versus the frequency of the incoming radiation for sodium metal.

   a) What does the gradient of the line represent? **[1 mark]**

   b) Determine the gradient of the line and specify its units. **[2 marks]**

   c) The line is extended to cross the x-axis. What does this crossing point represent? **[1 mark]**

   d) The line is extended further to cross the y-axis. What does this crossing point represent? **[1 mark]**

   e) By extending the graph to cross the x-axis or otherwise, determine the work function of sodium. **[2 marks]**

   f) If radiation of wavelength 0.42 µm is incident on the surface of sodium metal, what is the maximum kinetic energy of the emitted photoelectrons? **[3 marks]**

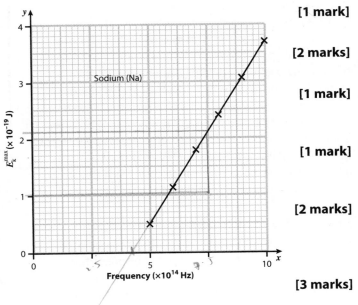

Energy Levels and Transitions

Ground States and Excited States

Electrons in atoms do not circulate around the nucleus in random orbits but do so in precise, well-defined orbits called **energy levels**. These are also referred to as **quantised energy levels** because they have fixed energy values associated with them.

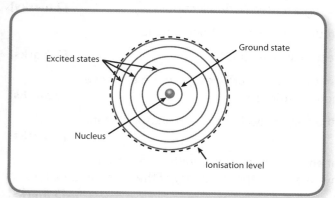

Taking the simplest case of a hydrogen atom, the electron in the lowest energy level is said to be in the **ground state**. If an electron gains energy then it may be possible to move to the next higher energy state or level, called the **first excited state**. If more energy is absorbed, electrons may be promoted to the **second excited state** or even higher energy states. To distinguish which levels or states electrons are in, each energy level is given a number, with $n = 1$ representing the ground state, $n = 2$ the next higher energy state and so on. As the value of n increases, the gap between energy states or levels decreases up to a level where the electron can escape from the atom; this is called the **ionisation level**.

Energy Level Values

If a slice is taken through an atom then the energy levels can be thought of as a series of horizontal levels or states beginning with the ground state and ending with the ionisation level. Each level is assigned a number ($n = 1, 2, 3…$) and each level has a **unique and well-defined energy value**. These values may be expressed in joules or in electronvolts. The most bound electron is one in the ground state. Conventionally, a **negative sign** is attributed to all energy values to show the strength of the

attractive force that binds the electron to the atom, i.e. how much energy is required to remove an electron from that energy level. The ground state is therefore the most **negative energetic state**. In the simplest atom, hydrogen (composed of a single electron orbiting a nucleus of one proton), this has a value of -13.6 eV or -2.2×10^{-19} J; the next excited state has an energy value of -3.4 eV or -5.4×10^{-19} J and so on all the way to $n = \infty$, which is the **ionisation energy**. The ionisation energy of hydrogen is 13.6 eV, which is the amount of energy that is needed to completely remove the electron from the lowest energy state of the hydrogen atom. When an atom loses (or gains) an electron, it will have a net positive (or negative) charge and it is called an **ion**.

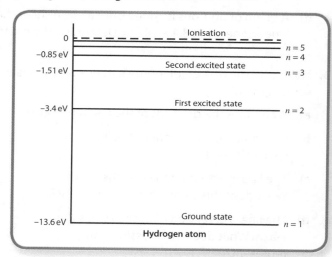

Excitation Processes

In order for an electron to be placed in an **excited state**, it must absorb exactly the right amount of energy to be promoted from one energy level to the next. This can happen in one of two ways:

- ● by the absorption of an incoming **photon** with the **exact amount of energy** to bridge the gap between any two energy levels

- ● through the interaction or collision with an incoming **electron** that has an energy equal to or greater than the energy required. The incoming electron loses energy to the atomic electron, which is then in an excited state; any excess energy remains with the incoming electron, which may leave the atom after the collision with much lower speed (less kinetic energy).

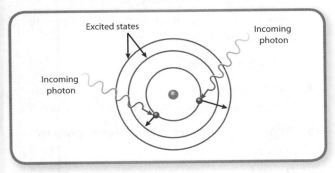

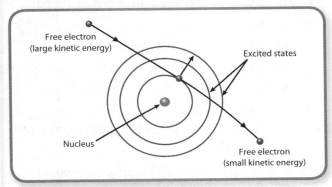

The key transitions in a hydrogen atom are shown. Transitions from higher energy states to the ground state give rise to a series of photon emissions called the **Lyman Series**. The energies and corresponding wavelengths associated with these transitions lie in the **ultraviolet (UV)** range. It is only the transitions from higher energy states to the first excited state, such as 6→2, 5→2, 4→2 and 3→2, that give photon energies and hence wavelengths in the **visible spectrum**, and this is referred to as the **Balmer Series**. The other major series is called the **Paschen Series** and these emissions appear in the **infrared (IR)** range.

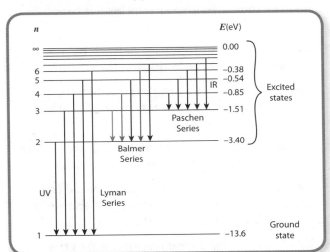

Photon Emission and De-excitation

An electron (or electrons) in an excited level is in an unstable state and drops down from its higher energy level to a lower energy level in a process called **de-excitation**. This process always results in the emission of a **photon**. The precise energy difference between levels (it may drop down several levels in one go) defines the precise energy of the emitted photon. Discrete energy levels therefore give rise to discrete emitted photon energies.

If the energy of the ground state ($n = 1$) is represented by E_1 and the first excited state ($n = 2$) by E_2 then the energy of the **transition** between $n = 2$ and $n = 1$ is given by:

$$\Delta E = E_2 - E_1$$

The frequency or wavelength of the emitted photon can then be determined by:

$$\Delta E = hf = \frac{hc}{\lambda}$$

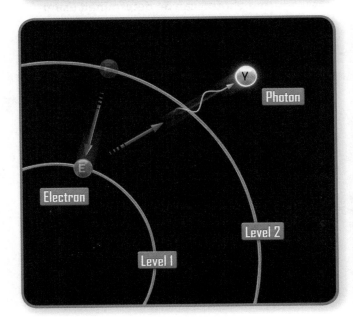

SUMMARY

- **Energy levels** in an atom are not continuous but discrete or **quantised** and each energy level is given a number to represent its state; the **ground state** is $n = 1$ and the **first excited state** is $n = 2$ and so on

- The energy values of each level are **negative** because the electrons are bound in the atom

- The most negative energy level is the **ground state**

- The **zero energy level** is defined by $n = \infty$ and is called the **ionisation level**

- Electrons can be promoted to higher energy levels by **photon absorption** or **electron collision**

- Electrons in excited states are unstable and **de-excite** with the **emission of photons**

- The **transition energy** is determined from the difference between the two energy levels through the equation $\Delta E = E_2 - E_1$

- The **frequency** (and wavelength) of the **emitted photon** can be determined using the equation $\Delta E = hf = \frac{hc}{\lambda}$

- In the hydrogen atom, the transitions between excited states and the ground state give photon emissions known collectively as the **Lyman Series**, and these occur in the **UV**

- Transitions between higher energy levels and the first excited state ($n = 2$) gives rise to the **Balmer Series**, and these occur in the **visible region**

- Transitions to the second excited state ($n = 3$) from higher energy levels gives a series in the **IR** called the **Paschen Series**

QUICK TEST

1. What is meant by the term 'ionisation energy'?

2. What is the name given to the lowest energy state for electrons in an atom?

3. What is 'excitation'?

4. What happens when an electron moves from a higher energy level to a lower energy level?

5. Which number energy level corresponds to zero energy?

6. How is the energy of a transition calculated?

7. In which part of the spectrum would you expect the Balmer transitions to occur?

8. What happens to an electron when it absorbs a photon of exactly the right energy?

9. An electron moves from $n = 3$ (-1.51 eV) to $n = 2$ (-3.40 eV). Determine the wavelength of the emitted photon.

10. In which part of the visible spectrum would the above transition occur?

11. An electron may be excited in one of two ways. State the methods by which this can occur.

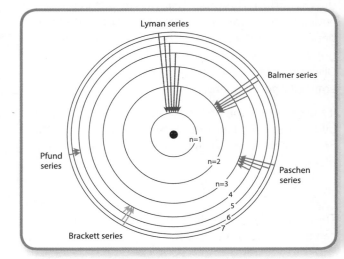

PRACTICE QUESTIONS

1. An electron is accelerated through a potential difference of 12.5 volts.

a) Calculate the gain in kinetic energy in (i) electronvolts and (ii) joules. **[2 marks]**

This electron then collides with a hydrogen atom.

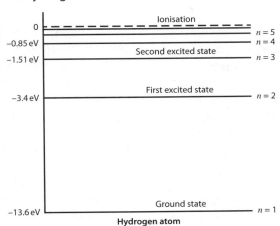

b) Using the energy level diagram shown, describe what happens within the atom. **[3 marks]**

c) Determine the three energy transitions that are possible when the excited electron drops down to its lowest energy state. **[1 mark]**

d) Calculate the energies of these three transitions in eV. **[3 marks]**

2. The following diagram shows the energy states in lithium metal.

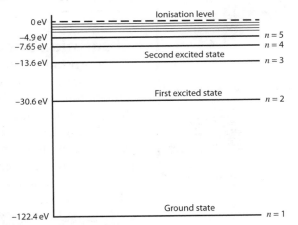

a) Express the ground state energy in terms of joules. **[1 mark]**

b) What is the ionisation energy of lithium? **[1 mark]**

c) An electron is in the excited state $n = 3$.

  (i) Write down the possible transitions from this level to the ground state. **[1 mark]**

  (ii) Determine the photon energies emitted in these processes. **[3 marks]**

d) Calculate the wavelength of the transition between $n = 5$ and $n = 4$ and state whether the photon is emitted in the IR, visible or UV regions of the spectrum. **[3 marks]**

Emission and Absorption Spectra

Continuous Spectra

Light from the Sun is called **white light** as it is composed of all colours of the **visible spectrum**. When passed through a prism, white light forms a continuous spread of merged colours between 400 nm and 700 nm, called a **continuous spectrum**. Light from a **filament bulb** (or incandescent bulb) also gives rise to a continuous spectrum. When a filament is heated, the energies given to the electrons promotes them to excited states and these de-excite to give photon emission. However, these **energy levels overlap**, allowing all possible photon emissions and thus a continuous spectrum of colours is observed.

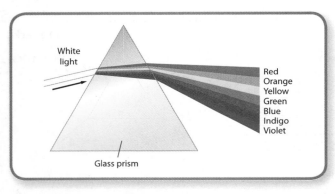

White light
Red
Orange
Yellow
Green
Blue
Indigo
Violet
Glass prism

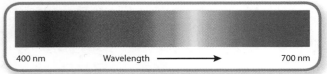

400 nm   Wavelength ⟶   700 nm

Emission Line Spectra

However, when light from a **fluorescent tube** is passed through a prism (or, better still, a **diffraction grating**), a spectrum of discrete bright coloured lines on a dark background is produced. This type of pattern is called a **line spectrum**. The spectrum observed from a fluorescent tube is similar to the Balmer lines observed in the spectrum of hydrogen; both provide clear evidence that electrons in atoms exist in **discrete energy levels**. When electrons

receive energy, for example by heating or through photon absorption/electron collision, they become **excited** into higher energy levels. When they return to the **ground state** (or any lower energy level), they emit photons of energy equal to the difference between the two energy levels. If these photons are in the visible region of the spectrum, they correspond to the lines seen in the spectrum; as no other energies are emitted, the background appears black. This is known as a **line emission spectrum**. Other examples include the colours seen when various compounds are heated in a Bunsen burner flame, with the colours resulting from the emission of photons due to energy level transitions, and light from astronomical gas clouds such as Orion Nebula.

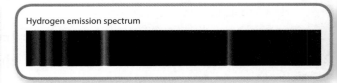

Hydrogen emission spectrum

Absorption Spectra

An **absorption spectrum** is obtained when white light passes through a **cool gas**. At these low temperatures, most of the electrons in the gas are in the ground state (lowest energy level). Photons of specific wavelengths are absorbed by the electrons as they are excited to higher levels. All other photons pass through as they cannot be absorbed.

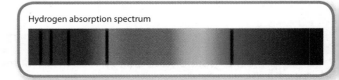

Hydrogen absorption spectrum

The absorption spectrum obtained therefore shows a **continuous spectrum** (i.e. photons that have passed straight through) but with **dark lines** indicating the energy (wavelength) of the absorbed photons. Although the excited electrons fall back to their original energy state and therefore emit photons in the process, these are **emitted in all directions** and hence the spectrum retains its dark lines. Because both emission and absorption involve the same energy levels, both types of spectra contain lines at identical positions and these lines are unique to the element.

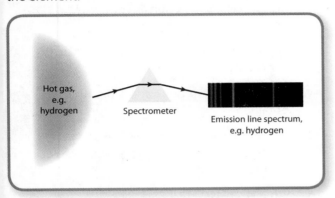

Emission line spectrum, e.g. hydrogen

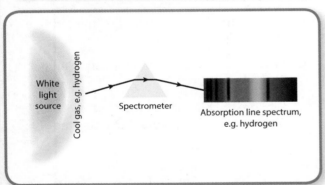

Absorption line spectrum, e.g. hydrogen

Spectral analysis is a technique used extensively to identify elements and it is particularly useful in studying the composition of stars and galaxies and other **astronomical objects**. An element can be identified by the unique absorption spectrum it produces. The absorption spectrum of the Sun, for example, reveals over 700 dark absorption lines, known as **Fraunhofer lines**, from elements within the Sun's structure.

The Fluorescent Tube
One type of energy-saving lighting makes use of **fluorescent tubes**. These long, thin glass tubes are filled with **mercury vapour** and coated inside with a fluorescent material such as phosphor. Applying a voltage across the tube provides heat to the **cathode**, which generates fast-moving electrons by **thermionic emission**. As these electrons travel towards the anode,

they collide with electrons in the mercury vapour and excite them into higher energy states; they even ionise the mercury atoms, creating a mixture of ions and free electrons called a **plasma**.

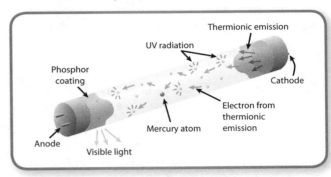

Electrons in excited states subsequently fall back to the ground state, releasing photons in the **UV range**. These photons interact with and are absorbed by the **phosphor coating** lining the tube. This absorption of energy promotes the electrons in the phosphor to higher energy levels and it is these that de-excite back to the ground state by emitting photons in the **visible light** part of the spectrum. The tube also gives out energy in the form of **heat**.

SUMMARY

- **White light consists of a continuous range of wavelengths (or frequencies) within the visible region that gives rise to a continuous spectrum**

- **A line spectrum is a series of discrete coloured or dark lines on a dark or bright background, respectively, that is characteristic of the atoms in a gas**

- **An emission line spectrum is a series of bright lines on a dark background that correspond to the wavelengths of the emitted photons when excited atoms return to their ground state**

- **An absorption line spectrum is a spectrum of dark lines on a coloured (continuous) spectral background produced when a gas absorbs photons from light that passes through it**

- Comparison of emission and absorption spectra of the same gas shows lines at identical positions indicating identical energy levels

- The Sun's absorption spectrum consists of hundreds of absorption lines called Fraunhofer lines

- Prisms and diffraction gratings can be used as spectrometers to obtain emission and absorption spectra

- Thermionic emission is the result of a heated cathode (filament) emitting fast-moving electrons

- Fluorescence is when a material absorbs short-wavelength (e.g. UV) radiation and re-emits it at a longer wavelength (e.g. visible light)

- A fluorescent tube is an energy-saving lighting device that emits visible light when its inner phosphor coating fluoresces as a result of the absorption of UV light

- A plasma is a mixture of electrons and ions in a gas

QUICK TEST

1. What is a line emission spectrum?

2. What spectrum is obtained where all frequencies of radiation are present?

3. Name two devices that allow radiation (light) to be dispersed or diffracted into its component colours.

4. How can an absorption spectrum be obtained?

5. Why are the lines in an emission and absorption spectrum for a particular gas at identical positions?

6. What is a plasma?

7. What is the process called when the cathode is heated to liberate electrons?

8. What are the two key components of a fluorescent tube?

9. Why are the photons emitted in the fluorescent tube important to obtaining visible light?

10. What area of physics uses absorption spectra in characterising the components of a gas?

PRACTICE QUESTIONS

1. Fluorescent tubes are used extensively around the world. They are tubes filled with mercury vapour at low pressure and give off light in the visible region.

   a) Explain how mercury atoms become excited in a fluorescent tube and emit photons. **[4 marks]**

   b) What is meant by the term 'fluorescence'? **[2 marks]**

   c) Explain how visible light is produced in a fluorescent tube. **[3 marks]**

   The quality of your written communication will be assessed in this question.

2. The diagram shows part of the emission spectrum for helium. The lowest energy level in a helium atom (the ground state) is −24.6 eV and there are a considerable number of excited states above this.

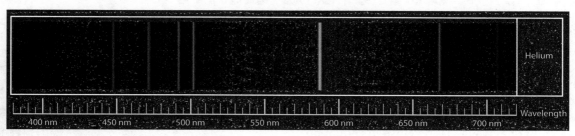

a) What is meant by an emission spectrum? **[2 marks]**

b) What does the ground state signify? **[1 mark]**

c) Owing to the complex nature of helium, there are two possible transitions from level 3 to level 2, with energy differences of (i) $\Delta E = 1.86$ eV and (ii) $\Delta E = 2.79$ eV. Calculate the wavelengths of the emitted photons in nm in each case and identify their colour using the emission spectrum shown. **[2 marks]**

d) Determine the wavelength for the transition from $n = 2$ (−4.4 eV) to the ground state and indicate what part of the spectrum it would appear in. **[3 marks]**

3. The Sun's absorption spectrum shows over 700 dark absorption lines (Fraunhofer lines) on top of a continuous spectrum between 300 nm and 900 nm.

a) With reference to the Sun's cooler atmosphere, explain why these dark lines appear and what they represent. **[3 marks]**

One of these dark lines corresponds to what is a prominent feature of the hydrogen spectrum, the H_α line at a wavelength of 658 nm.

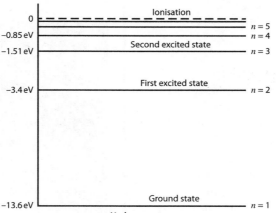

b) Using the energy level diagram shown, calculate the wavelength of the photon emission for the transition between levels $n = 3$ to $n = 2$ and compare this with the figure given above for H_α. **[3 marks]**

c) In which part of the visible spectrum does this line appear? **[1 mark]**

Wave–Particle Duality

De Broglie Wavelength

Louis de Broglie in 1923 reconciled the corpuscular or particle theory and the wave theory of light by proposing a wavelength of a wave that is associated with a moving particle. As electromagnetic waves could be thought of as particles, then particles could also be thought of as waves. Using Einstein's equation $E = mc^2$ and Planck's equation $E = hf$, de Broglie was able to express the **momentum of a photon** in terms of the photon's wavelength λ through the equation:

$$\lambda = \frac{h}{p}$$

where h is Planck's constant and p is the momentum of the photon. Moreover, de Broglie proposed that not just photons but other particles, such as electrons, would also show a **de Broglie wavelength** according to the equation:

$$\lambda = \frac{h}{p} = \frac{h}{mv}$$

where v is the velocity of the particle and m its mass. An electron travelling at, say, 0.1% of the speed of light would therefore have a wavelength of 2.4 nm. This equation shows that the faster a particle is moving the shorter is its corresponding wavelength.

Electron Diffraction

Both **diffraction and interference** patterns with light were already known and the process of producing alternating bands of light and dark through constructive and destructive interference was well established and fully explained using wave theory. Based on the de Broglie hypothesis, it should be possible to produce similar patterns using electrons. It was not until 1927 that **Davisson and Germer** produced interference rings using **electrons diffracted** through crystals in the same way that **X-rays** had been used through similar crystals much earlier. This was because the de Broglie wavelength of electrons travelling at about 1% of the speed of light have wavelengths comparable with those of X-rays. The results confirmed that electrons showed **wave-like properties**.

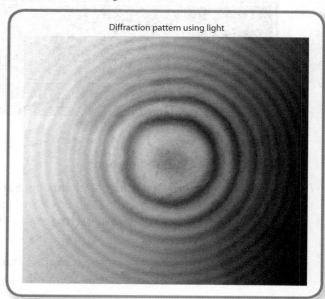

Diffraction pattern using light

Electrons that are accelerated with a smaller accelerating voltage, and hence have lower speed and longer wavelength, give rise to more **widely spaced rings** or **diffraction patterns**, in accordance with the wave theory of diffraction.

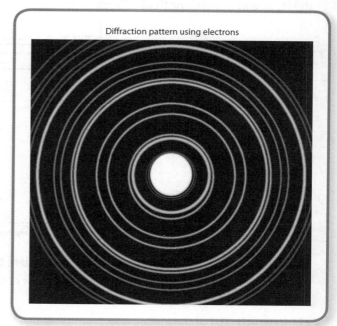

Diffraction pattern using electrons

Imaging with Electrons

Diffraction effects are only significant when a particle interacts with an object that is of **similar size** to its de Broglie wavelength. A shorter de Broglie wavelength will give rise to less significant diffraction and hence a clearer, better resolved image. An **electron microscope** utilises this effect and can provide much more image detail than an ordinary **light microscope**.

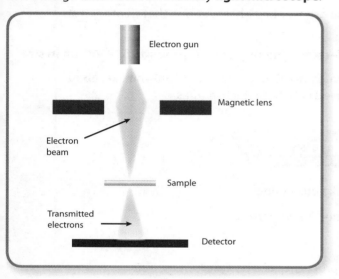

Electron gun

Magnetic lens

Electron beam

Sample

Transmitted electrons

Detector

Wave–Particle Duality

The idea that particles such as electrons, protons … and light (i.e. all electromagnetic radiation) can behave as both waves and as particles is known as **wave–particle duality**. Two key experiments demonstrated wave–particle duality very clearly:

● the photoelectric effect

● electron diffraction.

An explanation of the **photoelectric effect** was put forward by Planck based on his quantum theory but it was Einstein who firmly established the theoretical foundation that fully explained the emission of photoelectrons above the threshold frequency.

Light had previously been thought of as **waves** but both Planck and Einstein showed that light also behaved as a small, indivisible package or **quantum** of energy given by the equation $E = hf$, where f is the frequency of light in Hz. In the photoelectric effect, the incident quantum of light, now thought of as a particle, gives up its entire energy in the collision with a surface electron. Depending on the frequency (and thus energy) of the photons, it is possible using **Einstein's photoelectric equation** to determine the nature of the emission.

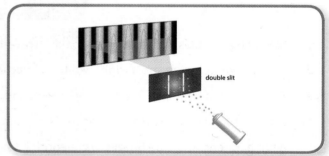

double slit

Electron diffraction proved that not only do electrons diffract but they also produce interference effects akin to those of light waves. This duality of the nature of particles eventually gave rise to a **quantum theory** that completely altered our way of thinking about the microscopic world. In this new world, matter and radiation could be described sometimes using a wave model and sometimes using a particle model.

SUMMARY

● **Light waves exhibit both** diffraction **and** interference **effects**

● **Diffraction** of electrons produces alternating light and dark bands due to interference effects between the electrons, similar to those of light

● **Electron diffraction** provides evidence that particles also behave as waves

● **The wavelength of a particle is given by its** de Broglie wavelength **that connects Planck's constant to the particle's momentum (mass and velocity) via the equation**

$$\lambda = \frac{h}{p} = \frac{h}{mv}$$

● **Crystal diffraction** requires electrons with very small de Broglie wavelengths, **hence fast speeds, to enhance the diffraction effect**

● **Electron microscopy** requires electrons with even smaller de Broglie wavelengths **in order to minimise diffraction and provide high quality images**

● **Wave–particle duality** is a notion that matter and radiation sometimes behave as particles and sometime behave as waves

● **The** photoelectric effect **and** electron diffraction **are two key observations that demonstrate wave–particle duality**

● **Light microscopes are limited by diffraction effects for imaging objects around 200–500 nm in size**

● **Electron microscopes are diffraction-limited for imaging objects of around 0.004 nm, i.e. about 100 000 times smaller than those that can be viewed with light microscopes**

QUICK TEST

1. Give two examples that show light to have wave-like properties.

2. Give an example that shows light exhibiting particle-like properties.

3. What is meant by the term 'de Broglie wavelength'?

4. The de Broglie wavelength is connected to which properties of a particle?

5. What happens to the de Broglie wavelength if the speed of the particle is increased?

6. If the de Broglie wavelength becomes longer, what is the effect on the energy of the particle?

7. What is meant by 'wave–particle duality'?

8. Give the expression that connects energy with wavelength.

9. Why was the photoelectric effect a key experiment to explain wave–particle duality?

10. What is the effect on the diffraction pattern if electrons with a greater speed are used?

11. Give an example of the application of fast-moving electrons acting like waves.

12. Which of the following can show diffraction effects: (i) neutrons (ii) microwaves (iii) molecular beams?

PRACTICE QUESTIONS

1. Electrons can exhibit wave-like properties.

   a) Explain what is meant by the term 'wave–particle duality'. **[2 marks]**

   b) Electrons are travelling at a speed of $5.6 \times 10^5 \, \text{m s}^{-1}$. Calculate the de Broglie wavelength of these electrons (electron mass is 9.11×10^{-31} kg). **[2 marks]**

   c) Calculate the speed of a proton that has the same de Broglie wavelength as the electrons (proton mass is 1.67×10^{-27} kg). **[2 marks]**

   d) Which of these particles would be more suited to the investigation of the structure of materials? State your reasons for this choice. **[2 marks]**

2. The Large Hadron Collider provides beams of high-energy protons that circulate with a speed that is 99.999 999 99% of the speed of light.

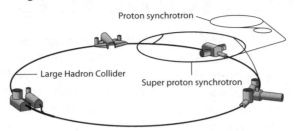

a) Calculate the de Broglie wavelength of the protons (proton mass is 1.67×10^{-27} kg; speed of light is 3.0×10^8 m s^{-1}) **[2 marks]**

b) Express this wavelength in terms of femtometres (ignore relativistic effects). **[1 mark]**

c) If relativistic effects are included, the wavelength calculated in part **b)** is ≈ 4 orders of magnitude smaller. What structures can be probed using protons of this energy? Give a reason. **[2 marks]**

d) Explain whether diffraction effects would be significant or not. **[2 marks]**

3. A bunch of electrons is accelerated in an electron microscope through a potential difference of 100 kV.

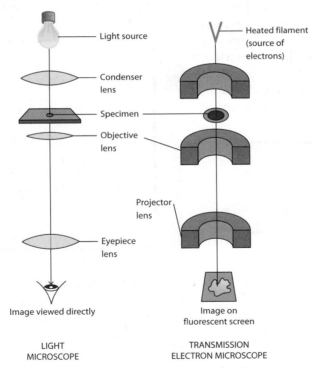

a) Calculate the kinetic energy in joules of each electron in the beam, assuming no energy losses. **[2 marks]**

b) Determine the speed of the electron beam (ignore relativistic effects). **[3 marks]**

c) Calculate the de Broglie wavelength of the electron beam. **[2 marks]**

d) What is the ratio of this wavelength compared with that of the light used in a light microscope that has a mean wavelength of 500 nm? **[1 mark]**

Waves and Vibrations

Wave Types

Waves represent the way **vibrations** are transmitted and they are divided into two types, **longitudinal waves** and **transverse waves**. Longitudinal waves vibrate **parallel** to their direction of travel. Examples include **sound waves**, **primary seismic waves (P-waves)** and **compression waves**, as can be demonstrated by a Slinky toy being pushed and pulled horizontally. Zones of denser vibrations (**compression**) are seen together with zones of less denser vibrations (**rarefaction**) moving along as the wave travels. In contrast, **transverse waves** have vibrations that are **perpendicular** to the direction of travel. Examples include **electromagnetic waves**, **secondary seismic waves (S-waves)** and waves on a **string**. Transverse waves can also be demonstrated with a Slinky toy. Waves that require a medium to travel through are called **mechanical waves**. Electromagnetic waves oscillate **electric** and **magnetic fields** that are at right angles to each other and the electromagnetic spectrum extends from radio waves through to visible light and beyond to X-ray and gamma radiation.

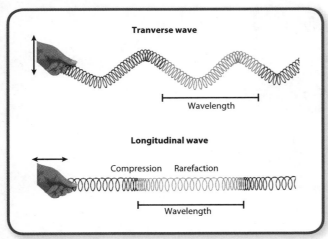

Tranverse wave

Wavelength

Longitudinal wave

Compression   Rarefaction

Wavelength

Describing Waves

All waves can be represented by a **sinusoidal curve**. The difference between a point on a wave and its equilibrium position is called its **displacement** and the maximum displacement is known as the **amplitude**. The distance between two corresponding points on the wave defines its **wavelength**, λ, and the time for one complete wave to pass a fixed point is referred to as the **period**. For waves with a period T, the frequency is given by $f = \frac{1}{T}$. The higher the frequency of a wave, the shorter the wavelength. The speed of a wave, the **wave speed**, is constant, and given by the distance travelled in one cycle divided by the time taken to complete one cycle, i.e. $v = \frac{\lambda}{T} = \frac{\lambda}{1/f} = f\lambda$. Wave speeds can vary for longitudinal waves; for example, sound travels at different speeds in different materials: $\sim 330 \, \mathrm{m\,s^{-1}}$ in air, $\sim 1200 \, \mathrm{m\,s^{-1}}$ in water and $\sim 3000 \, \mathrm{m\,s^{-1}}$ in a metal. Wave speeds can also vary for transverse waves; however, electromagnetic waves only travel at the same speed in a vacuum, and this is given by the speed of light in free space, i.e. $3.0 \times 10^8 \, \mathrm{m\,s^{-1}}$ – the symbol 'c' is always used.

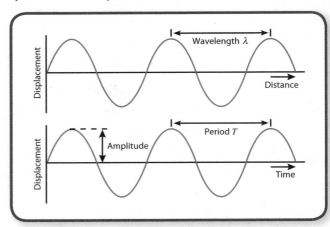

Wavelength λ

Displacement

Distance

Amplitude

Period T

Displacement

Time

Wave Polarisation

Another property of waves is the degree to which they can be **polarised**, i.e. the planes of vibration (when the cross-section of the wave is considered restricted). Longitudinal waves cannot be polarised but transverse waves can. Visible light from a filament lamp or the Sun is said to be **unpolarised**, i.e. the vibrations take place in all planes. Transverse waves can be polarised if the vibrations are restricted to one particular plane, and the wave is then said to be **plane-polarised**. Light reflected from a surface exhibits **partial polarisation**.

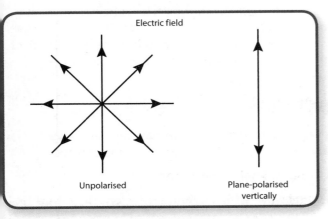

Electric field

Unpolarised

Plane-polarised
vertically

Only the oscillations associated with the electric field are considered. Polarising filters can be used to plane-polarise a transverse wave or can be used to 'block' the wave completely by using two polarising filters perpendicular to one another. For example, **Polaroid sunglasses** reduce the amount of sunlight received through polarisation effects. **Radio waves** are emitted from transmitters as polarised waves and thus radio and television receivers have to be orientated in a particular plane to ensure that they receive the polarised signals.

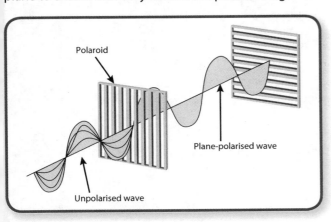

Polaroid

Plane-polarised wave

Unpolarised wave

Phase Difference

When waves of a particular wavelength and frequency are emitted, some waves may **lag** behind others and they are said to have a **phase difference**. Phase difference is the fraction of a cycle between two waves and is measured in degrees or radians, where 1 cycle = 360° = 2π radians. For two waves of wavelength λ, separated by a distance x, the phase difference in radians is given by $\frac{2\pi x}{\lambda}$ or $\frac{2\pi t}{T}$. Waves that have a **constant phase difference** produce two sets of **wavefronts** that interfere in a regular way, such as waves seen in a ripple tank. Wavefronts are lines perpendicular to the direction of travel of the wave

that are one wavelength apart. A phase change of 180° (π radians) often occurs when a wave is reflected at a boundary.

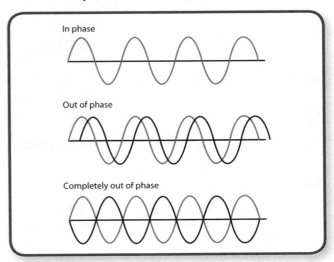

In phase

Out of phase

Completely out of phase

Wave Phenomena

When waves cross a boundary between two media at an angle there is usually a change in **wave speed** (and direction) accompanied by a change in wavelength. The effect is known as **refraction** and can easily be seen with wavefronts generated in a ripple tank. The **reflection** of wavefronts in which the wavelength remains the same can also be demonstrated using a ripple tank. Here wavefronts are incident upon an obstacle or obstacles to demonstrate wave properties such as reflection, refraction and interference. The way waves behave when they pass through a narrow gap or slit is called **diffraction**. Large gaps (much larger than the wavelength of the wave) result in little or no diffraction. Diffraction effects only become noticeable when the gap is of a similar size to the wavelength. In these cases, the **wavefronts** become more **spherical**

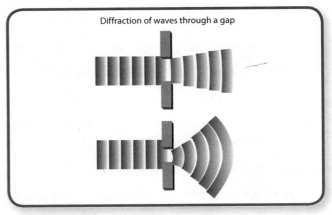

Diffraction of waves through a gap

and spread out in all directions beyond the gap. If the gap is smaller than the wavelength, there is greater diffraction. Similar effects are seen when wavefronts pass an obstacle, where diffraction effects are noticeable around the edges of the obstacle. Light waves can also be reflected, refracted and diffracted. The size of the gap must be of the order of magnitude of the wavelength of the light in order for diffraction to be observed.

SUMMARY

- There are two types of waves, longitudinal waves (e.g. sound waves and primary seismic waves) and transverse waves (waves on a string, secondary seismic waves and electromagnetic waves)

- Transverse waves can be plane-polarised whereas longitudinal waves cannot

- Waves can be described mathematically as a sinusoidal wave with a definite wavelength, period and amplitude

- The frequency of a wave is related to its period by $f = \frac{1}{T}$

- The wave speed of a wave connects the frequency and wavelength using the equation $v = f\lambda$. Electromagnetic waves are transverse waves that have a precise wave speed equal to the speed of light, i.e. $c = f\lambda$

- The phase difference between two waves is the fraction of a cycle the two vibrations are apart and is given by $\frac{2\pi x}{\lambda}$

QUICK TEST

1. What is the SI unit associated with the frequency of a wave?

2. What defines the amplitude of a wave?

3. Which two units are used to measure the phase difference?

4. Write down the equation that describes the wave speed for microwaves.

5. What is the connection between frequency and the period of a wave?

6. Describe what a longitudinal wave is.

7. Is the light from a lamp or the Sun polarised or unpolarised?

8. Give two examples of a transverse wave.

9. What happens to light when it is reflected from a surface?

10. Why do receiving TV aerials have to be orientated to receive signals?

11. What is meant by the 'plane of polarisation'?

12. What is meant by an unpolarised beam?

13. Sound waves with a frequency of 4000 Hz have a wavelength in air of 8.5×10^{-2} m. Calculate the wave speed of these sound waves.

14. What is the period of the sound waves in the above example?

15. What is the wavelength of red light if its frequency is 4.6×10^{14} Hz?

PRACTICE QUESTIONS

1. Earthquakes radiate two distinct types of waves, a transverse and a longitudinal wave.

   a) What names are attached to these two types of waves?   **[2 marks]**

   b) The diagram shows the displacement of rock in the Earth at a particular time for a transverse wave. What is the phase difference between the points A, B and C with respect to the origin?   **[3 marks]**

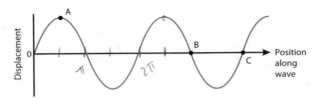

   c) The frequency of the seismic wave shown is 0.60 Hz and it travels at 4.8 km s^{-1}.

      (i) Calculate the wavelength of the wave.   **[2 marks]**

      (ii) Determine the period of the wave.   **[1 mark]**

   d) A longitudinal wave travels at 7.8 km s^{-1} and is emitted at the same time as the transverse wave. A seismologist detects that the two waves are 100 s apart. How far in km is the recording station from the epicentre of the earthquake?   **[2 marks]**

2. A microwave oven that is used to heat food emits microwaves with a frequency of 2.45 GHz.

   a) What type of wave is a microwave?   **[1 mark]**

   b) Describe how you would use a Slinky to demonstrate the nature of microwaves; draw a diagram to illustrate this movement and label the axes appropriately.   **[3 marks]**

   c) What happens to microwaves when they reflect from the walls of the microwave oven?   **[1 mark]**

   d) Calculate the wavelength of microwaves that have a frequency of 2.45 GHz.   **[2 marks]**

   e) Give another example of the use of microwaves.   **[1 mark]**

3. a) Define what is meant by a longitudinal wave.   **[1 mark]**

   b) Give two examples of a longitudinal wave.   **[2 marks]**

   c) Which of your examples in part **b)** are also mechanical waves? Give a reason for your choice.   **[2 marks]**

   d) How is the wavelength of a longitudinal waves defined?   **[1 mark]**

   e) Ultrasonic waves travel at 330 m s^{-1} in air but at 1.4 km s^{-1} in water. Compare the wavelengths of an ultrasonic wave of frequency 0.1 MHz in air and in water.   **[2 marks]**

Stationary Waves

Superposition of Waves

A wave moving along a string is known as a **progressive wave**. When a string on a musical instrument is plucked gently, the string vibrates and a sound is produced at a well-defined frequency. The vibrations set up in the string are reflected at both ends of the string and the two progressive waves meet and pass each other. The interaction between the waves is known as **superposition**, and the **displacement** at any point is the sum of the individual displacements of both waves at that point. This is called the **principle of superposition**. If the two opposing waves are in phase then reinforcement called **constructive interference** takes place; if the two waves are 180° or π radians out of phase then **destructive interference** occurs and the two waves cancel each other out.

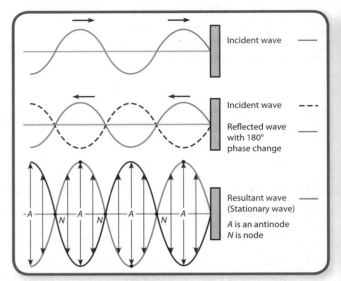

Incident wave ——

Incident wave – – –

Reflected wave with 180° phase change ——

Resultant wave (Stationary wave) ——

A is an antinode
N is node

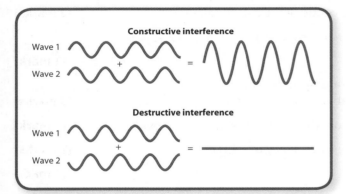

Constructive interference

Wave 1

Wave 2

+

=

Destructive interference

Wave 1

Wave 2

+

=

Formation of Stationary Waves

If two progressive waves of the **same frequency** (same wavelength) and at a **constant phase difference** (i.e. the two waves are **coherent**) pass each other, then cancellation and reinforcement may occur at several fixed positions along the length of the string. These effects are known as **interference**. For two coherent waves of equal amplitude travelling in opposite directions, the resulting pattern that it produces is known as a **stationary wave** (or **standing wave**). Stationary waves that vibrate freely do not transfer energy.

Points of no displacement, and hence zero energy, are called **nodes** and those of maximum displacement, or maximum energy, are called **antinodes**. Since the positions at the nodes are fixed, no energy transfer takes place.

Stationary Waves on Vibrating Strings

The simplest stationary wave pattern on a string produces the **fundamental mode of vibration** or the **first harmonic** (previously called the **fundamental**). It consists of a single loop with a node at either end and an antinode midway between. The distance between adjacent nodes, i.e. the length of the string, L, is then $L = \frac{1}{2}\lambda_1$, and the wavelength $\lambda_1 = 2L$. If the frequency increases then a second wave pattern emerges that has three nodes and two antinodes, and is called the **first overtone** or **second harmonic**. In this case, $\lambda_2 = L$. The next pattern produces a **third harmonic** where $\lambda_3 = \frac{2}{3}L$, and so on.

In terms of frequency, the fundamental frequency f_1 is given by $f_1 = \frac{v}{\lambda_1} = \frac{v}{2L}$, the frequency for the second harmonic is $f_2 = \frac{v}{\lambda_1} = \frac{v}{L} = 2f_1$ and for the third harmonic $f_3 = 3f_1$, and so on. In general, for any vibrating system with a node at either end, stationary wave patterns occur at frequencies f_1, $2f_1$, $3f_1$, …

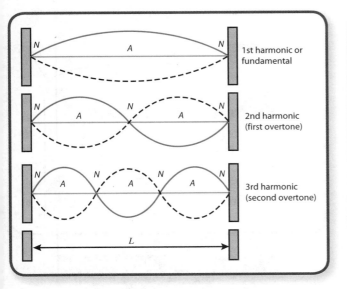

1st harmonic or fundamental

2nd harmonic (first overtone)

3rd harmonic (second overtone)

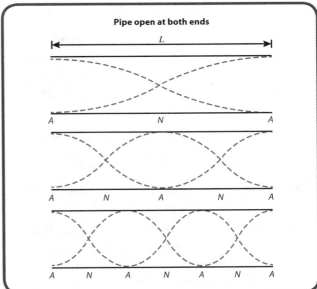

Pipe open at both ends

Stationary Waves in Air

Stationary waves can also be generated by sound in **pipes** or by using **microwaves**. Pipes closed at one end provide a node while the open end produces an antinode. In open pipes, antinodes are produced at both open ends. The resonant frequencies (i.e. the frequencies of the harmonics) depend on the length of the pipe.

The positions of nodes and antinodes can be effectively measured using a microwave emitter, a metal reflection plate and a detector as shown. The detector moving between the transmitter and plate provides a means of locating the nodes and antinodes and hence the wavelength of the microwaves involved. Other methods to demonstrate stationary waves include the use of lycopodium powder in a glass tube in resonance.

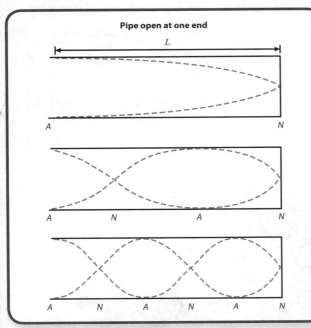

Pipe open at one end

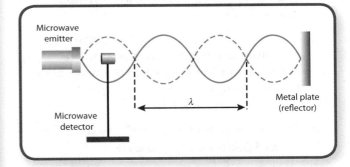

Strings under Tension

The **pitch** of a note is its frequency and musical instruments are 'tuned' to produce the correct fundamental pitch by altering the tension of the strings on the instrument. Comparisons are continuously made between the sound produced by a string under tension and a tuning fork of a precise frequency. A string is said to be 'tuned' if its fundamental frequency is identical with that produced by the tuning fork. The length of the string and the mass per unit length of the string are another two parameters that affect the pitch. The frequency or pitch is given by the equation $f = \frac{1}{2L}\sqrt{\frac{T}{\mu}}$, where L is the length of string or wire, T is the tension and μ is the mass per unit length of the string or wire. Increasing the tension or decreasing the length raises the pitch and, conversely, lowering the tension or increasing the length lowers the pitch.

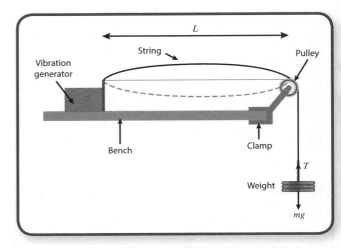

SUMMARY

- **Superposition occurs when two (or more) waves pass through each other**

- **The principle of superposition states that the resultant displacement of the two (or more) waves is the vector sum of their individual displacements**

- **Interference is the formation of a new wave pattern where two or more wave patterns overlap; for waves in phase, constructive interference results, while destructive interference results if waves are 180° out of phase**

- **To obtain interference, the two (or more) sources of waves must be coherent; coherent waves are waves of the same frequency (wavelength) with a constant phase difference**

- **Stationary waves (or standing waves) are produced when progressive waves are reflected at a boundary and the two waves travel through each other in opposite directions and interfere**

- **Stationary waves have nodes and antinodes at fixed positions along the wave**

- **The lowest possible resonant frequency is called the first harmonic or fundamental frequency**

- **Other resonant frequencies produced on a string have an exact number of half wavelengths that fit the length of the string and are called the second, third and higher order harmonics**

- **Stationary waves can be demonstrated using vibrating strings and microwaves**

QUICK TEST

1. How do stationary (standing) waves form?

2. What are the positions of zero and maximum displacement (amplitude) called?

3. What is meant by saying that the sources must be coherent to produce interference effects?

4. If two points on a wave have a phase difference of 1080°, are they in phase or out of phase?

5. What is the principle of superposition?

6. What effect results from the superposition of two waves in phase?

7. A stationary wave is produced on a stretched wire 0.8 m long that vibrates in its first harmonic mode. What is the wavelength of the stationary wave?

PRACTICE QUESTIONS

1. a) Explain what is meant by a stationary wave. [2 marks]

The frequency of microwaves generated on one side of a microwave oven is 2.45 GHz.

b) Calculate the wavelength of these microwaves. [2 marks]

c) When microwaves reflect from the sides of the oven, a stationary wave pattern is set up. With the aid of a diagram, give a simple illustration of this process, showing the positions of the nodes and antinodes. [2 marks]

d) In the above case, calculate the distance between nodes inside the oven. [1 mark]

e) If a bar of chocolate is placed in the microwave oven (no turntable used), what effect would be seen after about 30 seconds of irradiation? [2 marks]

2. A stationary sound wave is generated in a hollow pipe that has both ends open. The length of the pipe used is 1.2 m and the speed of sound in air is 330 m s^{-1}.

a) Draw two diagrams to illustrate (i) first and (ii) second harmonic of the modes of vibration in the open pipe; mark on your diagrams the positions of the nodes and antinodes. [3 marks]

b) Calculate the frequency of the first harmonic, the second harmonic and the third harmonic for this open system. [3 marks]

One end of the pipe is now closed.

c) Calculate the fundamental frequency and the first harmonic for this closed system. [2 marks]

d) Comment on the quality of sound resulting from both open and closed pipes. [2 marks]

3. The frequency f of waves that travel along a length of string or wire, of length L, under tension T, is given by:

$$f = \frac{1}{2L}\sqrt{\frac{T}{\mu}}$$

where μ is the mass per unit length of the strung or wire in kg m^{-1}.

a) What would happen to the frequency of a wave generated if (i) the tension in the string or wire is doubled or (ii) the length of the string is doubled? [2 marks]

The string is stretched between two points 0.65 m apart.

b) Calculate the frequency of the fundamental when the tension in the string is 100 N and the mass per unit length is 3.5 kg m^{-1}. [2 marks]

c) What would be the effect on the frequency if a lighter string is used? [2 marks]

Waves and Optics

Reflection and Refraction

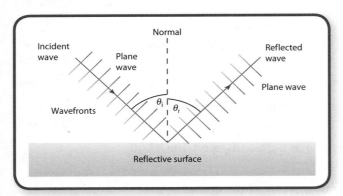

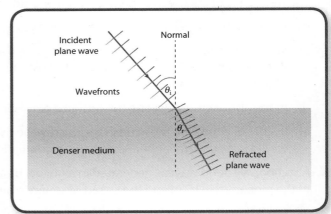

The **wave theory**, when applied to light, can be used to explain a number of optical phenomena such as reflection, refraction and internal reflection. It is usual to explain these effects using 'ray diagrams' where **light rays** represent the direction of travel of **wavefronts**. Reflection, for example, is the change of direction that occurs when light waves encounter a boundary between two media, such as air and a mirrored surface. The angle that the incident light wave makes to the **normal** (a line at right angles to the boundary surface) is called the **angle of incidence**, θ_i, and the **angle of reflection**, θ_r, is the angle of the reflected ray with the normal. The **laws of reflection** state:

● the incident ray, the reflected ray and the normal to the reflecting surface all lie in the same plane
● the angle of incidence equals the angle of reflection.

There is no change in wavelength of the light and therefore no change in the width of the wavefronts when reflection occurs. **Refraction**, however, is the change of direction that occurs when a light wave passes across the boundary between two transparent materials. When light travels from air into a more optically dense medium such as glass, the **angle of refraction**, θ_r, is always less than the **angle of incidence**, θ_i, i.e the light ray bends towards the normal in the glass. Wavefronts become closer together in a more optically dense medium; as the waves slow down their wavelength decreases. The ratio $\sin\theta_i / \sin\theta_r$ gives the **refractive index**, n, of the

more optically dense medium, in this case glass. The constancy of the ratio of the sine of the angles is called **Snell's Law**. For glass, the refractive index is 1.5, compared with the refractive index of air which is taken to be 1 (the actual value is 1.0003). When a light ray leaves the glass and enters the air, the light ray bends away from the normal so that light rays incident into the glass and leaving the glass are always parallel to each other.

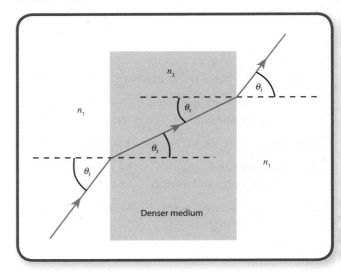

Refractive Index

Light rays entering a more optically dense transparent material interact with the atoms within it, resulting in the light ray slowing down to a speed considerably less than the speed of light in air (strictly, the speed of light in a vacuum). The **absolute refractive index** of a material is then the ratio of the speed of light in a vacuum (or air), c, to the speed of light in the material, c^*, and given by

$$n = \frac{c}{c^*} = \frac{\lambda_{\text{air}}}{\lambda_{\text{material}}}$$

Remember, the frequency of the light waves does not change when refraction takes place but the wavelength does.

For glass, $n = 1.5$, which means that the speed of light in glass is $2 \times 10^8 \, \text{m s}^{-1}$. If light rays pass from one material with refractive index n_1 into another material with refractive index n_2 then the **relative refractive index** between the two materials $_1n_2$ is the ratio of the speeds of light in both materials,

$$_1n_2 = \frac{c_1}{c_2} = \frac{n_2}{n_1}$$

This means that, while the absolute refractive index is a property of the material, the relative refractive index depends on the properties of the two materials involved in defining the boundary. Defining materials in this way allows **Snell's Law** to be given as

$$n_1 \sin\theta_1 = n_2 \sin\theta_2$$

where, θ_1 = angle of incidence in material 1 with refractive index n_1

θ_2 = angle of refraction in material 2 with refractive index n_2

Total Internal Reflection

When a light ray travels from, say, glass (a dense medium) into air (a less dense medium), it refracts away from the normal at the boundary (some light is, however, reflected and is this referred to as partial internal reflection). When the angle of incidence (within the denser medium) is increased up to a certain angle, the **critical angle**, θ_c, the refracted ray emerges along the boundary between the two materials. This means that the angle of refraction is equal to 90°. Applying Snell's Law gives $n_1 \sin\theta_c = n_2 \sin 90°$; hence $\sin\theta_c = \frac{n_2}{n_1}$. If n_2 is the refractive index of air then for glass–air refraction, $\sin\theta_c = \frac{1}{n_1}$, giving a critical angle for glass of 42°. If the angle of incidence increases further, i.e. exceeds the critical angle, then the refracted light ray is

totally internally reflected from the boundary layer between the two materials.

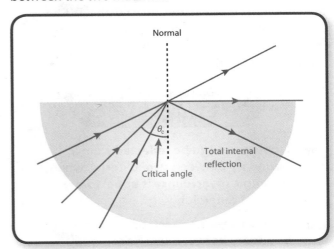

Optical Fibres

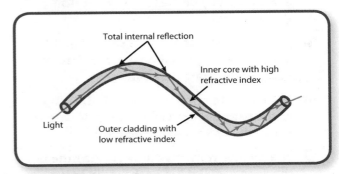

The phenomenon of total internal reflection has found numerous applications but most notably in the use of **optical fibres** for communication and in **medical endoscopes**. Optical fibres are made of very thin but flexible glass or plastic fibre surrounded by **cladding** that possesses a lower refractive index. The arrangement is such that any light ray hitting the boundary (between the glass and the cladding) does so at an angle greater than the critical angle so that all of the light is **totally internally reflected**. In communications, the core is usually extremely thin to reduce **multipath dispersion** (i.e. pulse spreading resulting from light taking different routes along the fibre), and **monochromatic waves** (waves of a single wavelength and frequency) can be used to avoid **spectral dispersion**, which also contributes to pulse broadening. Spectral dispersion arises because the speed of light is wavelength dependent when travelling in any optical medium other than a vacuum.

In medical endoscopes, two fibre bundles are used, one to provide illumination while the other is used to transmit the image of the object within the body cavity. This fibre bundle needs to be a **coherent bundle**, i.e. the fibres at each end have to be in the same relative positions.

SUMMARY

- Light waves can be **reflected** from a boundary between two material, where the angle of incidence is equal to the angle of reflection

- Light waves can be **refracted** at the boundary between two transparent materials, where the refracted ray is bent towards the normal when passing from a less optically dense to a more optically dense medium

- The ratio $\frac{\sin\theta_i}{\sin\theta_r}$ = constant is called **Snells's Law**; the ratio gives the relative refractive index of the more dense medium

- The **absolute refractive index** is the ratio of the speed of light in a vacuum (or air) to the speed of light in the medium

- The **relative refractive index** is the ratio of the speeds of light in both media or the ratio of their respective refractive indices

- For light travelling from a more optically dense medium to a less optically dense medium, if the angle of incidence is equal to the **critical angle**, the refracted ray is directed along the boundary between the two materials

- If the angle of incidence exceeds the critical angle, the ray is **totally internally reflected**

- Total internal reflection is used in **optical fibres** for communication and in medical endoscopes

QUICK TEST

1. What is the connection between a light ray and a wavefront?

2. What is meant by the term 'normal'?

3. In refraction, the frequency remains constant, but what varies?

4. What is 'partial internal reflection'?

5. Give the general equation for Snell's Law.

6. Define the absolute refractive index in terms of the speed of light.

7. What is meant by an 'optically dense' material?

8. Give an expression for the critical angle in terms of the refractive index for light passing from an optically dense medium to a less optically dense medium.

9. What happens to light rays when the angle of incidence exceeds the critical angle?

10. Give two applications where the above effect in question **9** is used.

11. What is an optical fibre?

12. In an optical fibre, where does the total internal reflection take place?

13. Why should the tube or core of optical fibres be very thin?

14. What is spectral dispersion?

PRACTICE QUESTIONS

Take the refractive index of air = 1.00, crown glass = 1.52 and water = 1.33; speed of light $c = 3.0 \times 10^8\,\text{m}\,\text{s}^{-1}$.

1. a) Define what is meant by the refractive index of a transparent medium. **[1 mark]**

    b) Calculate the speed of light in crown glass. **[2 marks]**

    c) A ray of light passes across a plane boundary between air and a rectangular block of crown glass at an angle of incidence of 55°. Show that the angle of refraction of this light ray is 33°. **[2 marks]**

    d) This refracted ray of light then enters water from the crown glass. What is the angle of refraction of the light ray as it enters the water? **[3 marks]**

    e) Explain the result obtained in part **d)**. **[2 marks]**

2. a) State two conditions for a light ray to undergo total internal reflection at a boundary between two optically transparent materials. **[2 marks]**

    b) Show that the critical angle at a boundary between crown glass and water is 61°. **[2 marks]**

    c) Explain what happens to the light ray in part **b)** if the angle of incidence is less than the critical angle. **[3 marks]**

    d) Calculate the angle of the refracted ray in part **c)** if the angle of incidence is 45°. **[2 marks]**

3. The diagram shows a section of a single fibre optic cable used as part of a bundle in a medical endoscope.

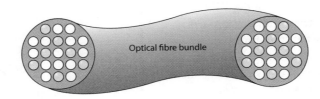

Optical fibre bundle

    a) Explain why the fibre bundle has to be coherent in order to view an image. **[1 mark]**

    b) What are the properties of the cladding used in optical fibres? **[2 marks]**

    c) If the refractive index of the core is 1.60 and that of the cladding is 1.50, calculate the critical angle for light to travel along the optical fibre. **[1 mark]**

    d) When white light is used, pulse broadening (spectral dispersion) can occur. Explain what is meant by pulse broadening and how it can be prevented. **[4 marks]**

Diffraction and Interference

Diffraction

Diffraction is the way that waves spread as they pass through a gap or pass by an edge. Diffraction is important in the design of **optical equipment**, such as microscopes and telescopes, that use small gaps to let light through. The diffraction of light (monochromatic, coherent light) through a **single opening** or **slit** results in a pattern of **light** and **dark fringes** or bands on a screen beyond the slit. The **central bright fringe** is twice as wide as each of the outer light fringes; the **outer fringes** are the same width as each other and are less intense than the central fringe, and the intensity decreases as the distance from the centre increases. The narrow dark fringes appear equally spaced apart on either side of the bright central fringe. The **narrower** the slit, the **broader** the diffraction pattern becomes. Increasing the wavelength of the light also broadens the pattern. The width of the central fringe, W is given by $W = \frac{2D\lambda}{a}$, where D is the distance from the slit and a is the slit width. The first minimum (dark band) occurs at an angle given by $\sin\theta = \frac{\lambda}{a}$ and, in general, the mth minimum is given by $\sin\theta = \frac{m\lambda}{a}$. Such diffraction patterns are also obtained by accelerated particles such as **electrons**, confirming the notion that particles also possess wave-like properties. Particle wavelengths are determined from their **de Broglie wavelength**.

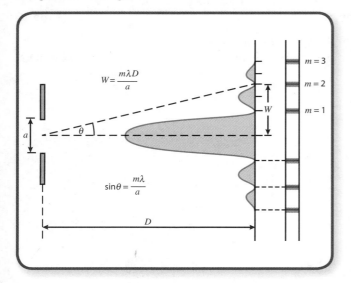

$$W = \frac{m\lambda D}{a}$$

$$\sin\theta = \frac{m\lambda}{a}$$

Double-slit Interference

When a light source is incident on two slits (**double slit**), not only is **diffraction** evident but also **interference**. The two slits act as a source of two waves of the same wavelength (frequency) and with a constant phase difference, i.e. **coherent sources**. **Double-slit interference** gives a series of equally spaced and alternate bright and dark **fringes** parallel to the slit, known as **Young's fringes**. The separation of the fringes, x, depends on the slit spacing, d, and on the distance between the slits and the screen, D, in accordance with the equation

$$x = \frac{D\lambda}{d}$$

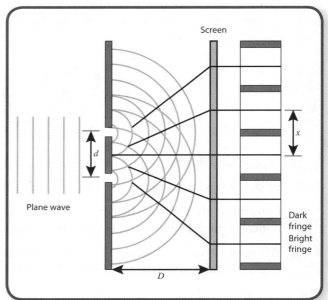

The intensity profile from double-slit interference is contained within the diffraction envelope from a single slit, as shown. The above expression can be derived from a geometric consideration of the differences in path lengths between the screen and the slits. If the path difference is a whole number of wavelengths then **constructive interference** is obtained and a **bright fringe** is seen; a path difference of a whole number of wavelengths plus half a wavelength leads to **destructive interference** and **dark fringes** are obtained.

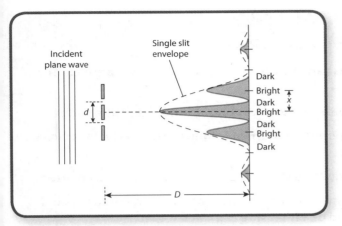

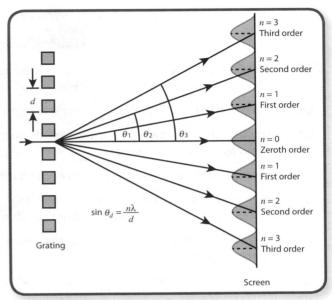

The fringe separation for say red laser light ($\lambda = 635\,\text{nm}$) is greater than that for blue light ($\lambda = 445\,\text{nm}$). Using a coherent white light source produces a white bright central fringe (all colours mixed) with outer white fringes that contain both extremes of red and blue fringes on their outer edges. Double-slit interference can also be demonstrated using sound waves, water waves or microwaves.

Diffraction Gratings

Diffraction gratings are simply an extension and replacement of the double slit by an array of **multiple slits**. The resulting **interference pattern** is essentially the same shape but, with so many beams reinforcing the pattern, the bright fringes are brighter and sharper and the dark fringes are darker and sharper. Because of this increased sharpness, measurements can be used to accurately determine the wavelength of light. The equation relating the wavelength to the slit separation, d, is

$$d \sin\theta = n\lambda$$

where n is called the **order of diffraction**. The slit separation or spacing of the grating is determined by $\frac{1.0 \times 10^{-3}}{n}$, where n is the number of grating lines per mm. The maximum brightness occurs at the centre and this is the **zeroth order** or $n = 0$. The first set of bright fringes on either side of this are called **first order** or $n = 1$, and the second set $n = 2$ or **second order**, and so on. The above equation can be derived from the geometry of light rays from adjacent slits.

The equation shows that:

● if λ increases then $\sin\theta$ increases and so θ is larger, i.e. the pattern will become more spread out
● if d increases then $\sin\theta$ decreases, i.e. a coarser grating produces a narrower pattern.

Diffraction Grating Using White Light

White light is composed of all the colours so an interference pattern produced by a diffraction grating produces a very colourful series with a sharp, bright and white zeroth order. The first order band is composed of the spectral colours from **blue to red** on either side of the **central white band**. This pattern continues for the second and third orders but the width of the spectral band increases as the order increases. A small spectrometer can be used to investigate interference patterns from different spectral sources. The maximum number of orders that may be seen is given by the value of $\frac{d}{\lambda}$ that is rounded down to the nearest whole number.

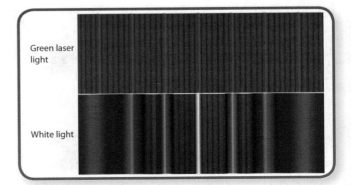

SUMMARY

- **Diffraction** is the spreading of waves as they pass an obstacle or through a narrow gap; it can also be observed with electrons and other particles
- A **diffraction pattern** is a bright central fringe with dark and less bright fringes on either side
- Diffraction through two slits produces an interference pattern of bright and dark equally spaced fringes called **Young's fringes**
- Sharper interference patterns are produced if multiple slits, called **diffraction gratings**, are used
- For white light as well as monochromatic light, **maximum brightness** occurs at the centre, called zeroth order ($n = 0$), with bright fringes produced on either side at $n = 1$, $n = 2$, etc.
- The equation relating the nth order, the angle, θ, and the wavelength, λ, for a grating with slits a distance d apart is $d \sin \theta = n\lambda$
- Interference and diffraction can be demonstrated with **light waves**, **microwaves** and **water waves**

QUICK TEST

1. Write down the fringe spacing for Young's double-slit experiment, defining the symbols used.

2. What is meant by coherent sources?

3. What is the main difference in the slit patterns between red and blue light?

4. When waves pass through a gap, the diffracted waves spread out more if: (i) the gap is narrower, (ii) the gap is wider, (iii) the wavelength is larger, (iv) the wavelength is smaller. Which two are correct?

5. Describe the overall shape of the diffraction of light through a single slit.

6. A diffraction grating has 1000 slits per mm. What is the spacing between the slits?

7. Monochromatic light is transmitted through a grating with 4.1×10^5 slits per metre. The first order maximum subtends an angle of 15°. What is the wavelength of the light used?

8. What are the key differences in the diffraction grating pattern produced by red laser light and white light?

PRACTICE QUESTIONS

1. A Young's double-slit experiment is set up in a laboratory using slits that are 0.3 mm apart. The slits are 2.8 m from the screen and a blue laser is used to produce the fringe pattern on the screen. The wavelength of the blue light is 445 nm.

   a) What does Young's double-slit experiment demonstrate? **[2 marks]**

   b) Describe the fringe pattern obtained. **[2 marks]**

   c) Calculate the fringe spacing. **[2 marks]**

   d) What would be the effect of replacing the blue laser with a red laser (the wavelength of the red laser is 635 nm)? **[1 mark]**

   e) Why are coherent light sources necessary for this experiment? **[2 marks]**

2. A diffraction grating has 650 slits per mm.

   a) Calculate the spacing of the slits on the grating. **[1 mark]**

   A red laser beam of wavelength 635 nm is directed onto the grating.

   b) Calculate the angles of the first and second order diffraction fringes using the laser light. **[2 marks]**

   c) Explain why it is not possible to observe the third order diffraction fringes. **[1 mark]**

   d) Describe the effect of using white light from a filament lamp through the same grating. **[2 marks]**

3. A narrow beam of white light is directed normally onto a single slit and a diffracted pattern is observed on a screen. The width of the central fringe is given by $W = \frac{2D\lambda}{a}$.

   a) Explain each of the symbols used in the above expression. **[4 marks]**

   b) Sketch a diagram that depicts the nature of the diffraction pattern in terms of the width and intensity of this central fringe. **[3 marks]**

   A blue filter is placed across the single slit and the diffraction pattern observed; the blue filter is then changed to a red filter.

   c) Describe the differences in the diffraction patterns with the blue and red filters compared with that that observed with white light in part **b)**. **[4 marks]**

   d) For the red filter, the measured distance across two fringes either side of the central fringe is 24 mm. Determine the width of the central fringe. **[2 marks]**

Scalars, Vectors and Moments

Scalar and Vector Quantities

A **scalar quantity** has no direction just size or magnitude; a **vector quantity** has both direction and size.

Scalar quantities	Vector quantities
mass, temperature, speed, distance, energy, time	weight, velocity, displacement, acceleration, momentum, force, moment

Vectors can be added together to give the **resultant**, R. This can be done either geometrically using a scale diagram (with a specified scale) and adding vectors 'tip-to-toe', or algebraically. Vectors can be added in any order, i.e. $a + b + c = c + a + b = b + c + a$. The resultant of two vectors at right angles to each other can be found using Pythagoras' Theorem, i.e. $R = \sqrt{a^2 + b^2}$ and $\theta = \tan^{-1}\left(\frac{b}{a}\right)$. For vectors not at right angles, the sine or cosine rules can be used.

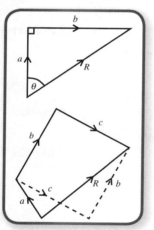

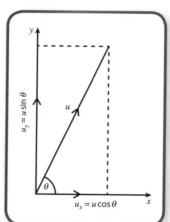

Any vector, u, can be resolved into two right-angle components; these may be along the x and y axes (as $u_x = u \cos \theta$ and $u_y = u \sin \theta$), or parallel and perpendicular to the slope on an inclined plane. Vectors are denoted by straight lines (indicating the size) with arrows (indicating the direction).

Forces in Equilibrium

The most common vector diagram is a **free body force diagram**, where all the forces acting on the object are drawn. (Note that the forces are usually all drawn so as to begin at the centre of mass of the object, though the vector representing friction may be drawn along the interface between the block and the surface.) An object is in **equilibrium** if all the forces acting on it are balanced and cancel each other out. The forces can each be resolved into vertical and horizontal components (or components perpendicular and parallel to the slope) to solve equilibrium problems. In a tip-to-toe vector diagram, three or more forces that are in equilibrium form a closed loop, (i.e. they start and finish in the same place) and hence have a zero resultant. Three forces that are in equilibrium would form a **closed triangle**. Forces include applied forces (P), weight (W) drag as well as contact forces such as friction (F). **Frictional forces** always oppose motion and frictional surfaces can be described as rough. The **weight**, W, of an object always acts vertically downwards whereas its **reaction force**, N, will act in a direction normal (90°) to the slope.

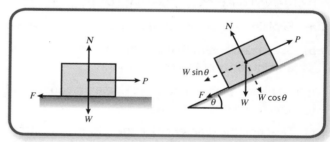

Moments

When a force acts about a particular point, called a **turning point** or **pivot point**, it causes a **turning effect** or **rotation** about that point; this turning effect is called a **moment** (the word **torque** is also used). The size of the moment, M, depends on the size of

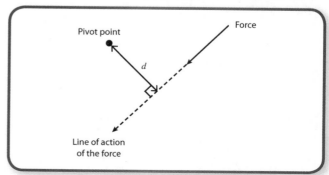

the force, F, and how far the force is applied from the turning point; this distance, d, is the perpendicular distance between the line of action of the force and the pivot point; then $M = F \times d$. Moments are vectors with units of $N\,m$.

Do not confuse the unit of moment of force with that of energy (the joule); for energy the product of the force and distance moved are along the same line.

The **principle of moments** states that, for an object to be in equilibrium (i.e. not rotating or accelerating), the **sum of the clockwise moments** must be equal to the **sum of the anticlockwise moments** about the same point. If this is not true then the object will rotate in the direction of the resultant moment.

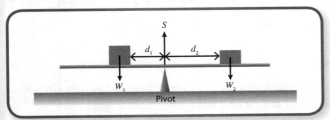

When an object is in equilibrium about a single pivot point, the downward forces must be balanced by an upward force acting through the pivot; this is called the **support force** or **reaction force**. If there are two pivot points then there are two support forces; these are in general not equal.

If an object is in equilibrium, two conditions must always apply:

- the vector sum of the all the forces must equal zero, i.e. any upward forces, e.g. reaction forces, must equal the downward forces, e.g. weight, i.e $S = W_1 + W_2$

- the sum of all the turning moments must equal zero, i.e. the sum of the anticlockwise moments, i.e. $W_1 \times d_1 = W_2 \times d_2$

Levers and Couples

A number of devices use the physics of moments to increase the force applied to an object. Examples include car jacks, spanners, crowbars and wheelbarrows. Increasing the distance d between the pivot point and the input force F increases the output force. Such force multipliers are referred to as **levers**.

A **couple** is a special case where a pair of equal forces act parallel to each other but in opposite directions. A couple may produce a turning force, called a **torque**, T, given by $T = F \times D$, where F is the magnitude of one of the forces and D is the distance between the two forces. Note that D is independent of the pivot point.

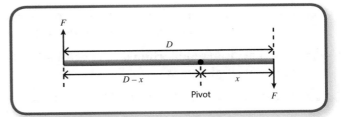

Stability and Toppling

The **centre of mass** of an object is the point through which a single force on the object results in no turning effect; it's the point through which the weight, W, of the object acts. An object will be more **stable** if it has a low centre of mass and a wide base, as the moment needed to turn the object is significantly greater.

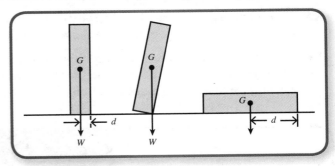

An object will **topple** over if the line of action of its weight falls outside the base area. The **critical angle** is when the object is on the point of toppling. When the line of action is outside the base, a resultant moment occurs that provides the turning force for the object to topple. The effect is more significant when objects are on slopes or inclined planes or are irregularly shaped.

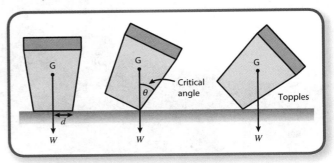

SUMMARY

- A **scalar quantity** is one which has size only; a **vector quantity** has both size and direction
- A **resultant vector** is a vector that arises from adding two or more vectors together; these may be forces, velocities or displacements, or any other vector quantity
- **Resolving a vector** means splitting a vector into two mutually perpendicular components that add up to the original vector; these can be added 'tip-to-toe'
- An object is in **equilibrium** if the vector sum of the forces is zero and the sum of the turning moments is also zero
- A **free body diagram** shows all forces acting on an object
- A **support force** or **reaction force** is a force that balances downward forces
- A **rough surface** means **frictional forces** are involved; a frictional force always opposes motion
- A **moment of a force** is the force multiplied by the perpendicular distance from the pivot; moments have units of N m and are vectors
- A **pivot** is a point about which an object may rotate
- The **centre of mass** of an object is the point through which a single force on the object results in no turning effect; it's the point through which the weight of the object acts
- A **couple** is a pair of equal and opposite parallel forces applied to the same body that do not act along the same line; it is equal to the force multiplied by the distance between the two equal forces
- The **critical angle** is when a tilted object is on the point of toppling; when the line of action through the centre of mass is outside the base, a resultant moment occurs that provides the turning force for the object to topple

1. A force of 8 N acts due north and another force of 6 N acts due east on an object. Draw a vector diagram and determine the size and direction of the resultant force.

2. A force of 24 N acts at an angle of 60° above the horizontal. Calculate the vertical and horizontal components of this force.

3. A box rests on a rough inclined plane at an angle θ to the horizontal. Draw a free body diagram to represent this situation. What vector diagram could be drawn to show that the box is in equilibrium?

4. A child of weight 250 N sits on one end of a 4.0 m long see-saw. For the see-saw to be in equilibrium, what distance from the centre would another child of weight 300 N have to sit? Ignoring the weight of the see-saw itself, what is the value of the support force above the pivot point?

5. A cyclist turns a corner and exerts two equal and opposite forces of 20 N on the handlebars that are 0.60 m apart. What is the size of the couple?

6. Why is it better for an object to have a low centre of mass?

7. What is meant by torque?

PRACTICE QUESTIONS

1. A piece of luggage is placed on a rough conveyor belt inclined at an angle of 30° to the horizontal as it is loaded onto an aeroplane. The weight of the luggage is 250 N and it is in equilibrium.

   a) Draw a force diagram to represent this situation. **[3 marks]**

   b) Resolve the weight of the box into its two components (one parallel to the plane and the other perpendicular to the plane). **[2 marks]**

   c) Draw a tip-to-toe vector diagram to represent this equilibrium condition. **[2 marks]**

2. A car of mass 1.2 tonnes is stationary on a road bridge that has a span of 25 m and is supported at either end by a concrete plinth. The bridge has a mass of 3.2 tonnes and its mass acts through its centre. The car is positioned 10 m from one end of the bridge and is in equilibrium (1 tonne = 1000 kg).

   a) Draw a force diagram to represent this situation showing the reaction forces at the pivot points and the forces on the bridge. **[3 marks]**

   b) Use the principle of moments to determine the reaction forces acting at each plinth. **[5 marks]**

   c) Explain what happens to the size of the reaction forces as the car moves to the centre of the bridge. **[1 mark]**

3. An empty wardrobe of mass 20 kg is 2.0 m tall with a width of 1.2 m. A horizontal force, F, is applied to the top edge of the wardrobe, causing it to tilt about its base at point X.

   a) Draw a diagram to show the forces acting, labelling the point X. **[2 marks]**

   b) If the wardrobe has a mass of 20 kg determine the force needed to tilt the wardrobe. **[3 marks]**

   c) Calculate the critical angle beyond which the wardrobe will topple. **[2 marks]**

   d) If the wardrobe contained some clothing items that lowered the centre of gravity to 0.5 m above the base, determine the new critical angle. **[2 marks]**

   e) What does this mean in terms of the force needed to tilt the wardrobe? **[1 mark]**

Motion in a Straight Line

Speed and Velocity

An object moving at a **constant speed**, v, travels equal distances, s, in equal intervals of time, t. The speed is given by $v = \frac{s}{t}$. If the speed is changing then the **average** or **mean speed** is $v = \frac{\text{total distance}}{\text{total time}}$. The **instantaneous speed** is the gradient of the line at that time on a **distance–time graph**. Speed is a scalar quantity (it only has size) with units of m s^{-1}. Velocity is a **vector quantity** (with both size and direction) and the value of a velocity along a line may be given with either a plus or minus sign to denote the direction. Velocities are found using displacements and not distances; **displacement** is a vector and gives the distance travelled in a given direction.

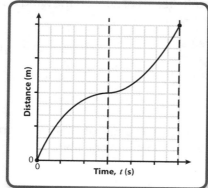

Displacement–time graphs allow the velocity of an object at any point to be found from the gradient of the line at that point. Differences

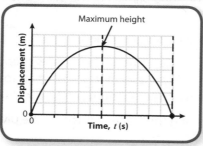

between distance–time and displacement–time graphs can be seen here, for the example of an object projected vertically upwards.

Acceleration

Acceleration is the rate of change of velocity per unit time; acceleration is also a **vector** and has units of m s^{-2}. **Uniform or constant acceleration** is where the velocity of an object changes at a constant rate and this will be a straight line on a **velocity–time graph**: for uniform acceleration the straight line will have a positive gradient whereas for uniform deceleration

it will have a negative gradient. **Non-uniform acceleration** is shown as a curve on a velocity–time graph and the instantaneous acceleration at any point is obtained by determining the gradient at that point. An object projected vertically upwards therefore has a velocity–time graph with a negative-gradient straight line illustrating constant acceleration of $-9.8\,\text{m s}^{-2}$. The **area** under a velocity–time graph gives the displacement.

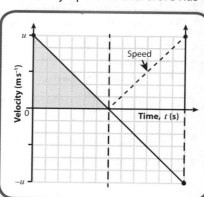

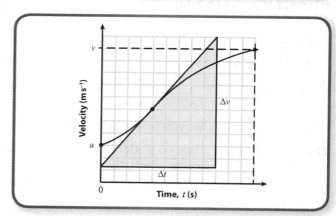

Equations of Motion

The following equations of motion, collectively called the **'suvat' equations**, can only be applied if the **acceleration is constant**. This includes the motion of objects under the influence of gravity near the Earth's surface, where the acceleration is that due to gravity, g.

$v = u + at$	$s = \dfrac{1}{2}(u + v)t$
$s = ut + \dfrac{1}{2}at^2$	$v^2 = u^2 + 2as$

The *suvat* equations can be derived using the following general velocity–time graph. When undertaking calculations, always begin by identifying which variables are known and which ones are needed; list all five parameters with any known values. Note the direction of the velocity, displacement and acceleration: some vectors may need to be given negative values (e.g. if the object is decelerating).

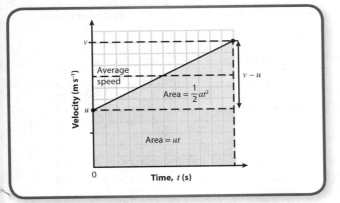

Free Fall

The position of an object that is dropped from a height can be calculated since the initial velocity is zero. The distance travelled (displacement) is given by $s = ut + \frac{1}{2}at^2$; with $u = 0$ this gives $s = \frac{1}{2}at^2$. If the displacement is plotted against t then a parabolic curve is obtained. However, plotting s against t^2 gives a straight line with a gradient of $\frac{1}{2}a$ or, in this case, $\frac{1}{2}g$, i.e. the acceleration due to gravity can be found. A similar experiment conducted on the Moon would give $g = 1.6\,\text{ms}^{-2}$.

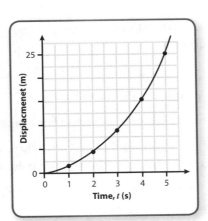

Two-Stage Motion Problems

There are many problems that require a solution involving two separate and distinct stages of calculation. For example, a car travelling at a constant speed before slowing down to a stop at traffic lights, or an object dropped and embedding itself into sand. In these circumstances, the problem is split into two separate stages with the solution from the first stage being used as the starting point for the second stage. Considering the ball dropped onto sand, the two stages are:

⬤ vertical motion under gravity, which is taken as constant, i.e. $a = g = 9.8\,\text{m s}^{-2}$

⬤ deceleration of the motion as it enters the sand before stopping in the sand.

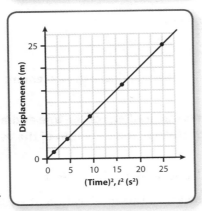

The acceleration in each stage is not the same. The final velocity determined in stage 1 is the initial velocity used in stage 2 to determine the deceleration and hence the distance travelled in the sand.

Newton's Laws of Motion

These laws describe the relationship between an object's motion and the forces that are acting upon it. **Newton's First Law** states that if an object has no resultant force acting on it then its velocity will not change, i.e. it is either stationary or moving in a straight line with a constant speed. If there are two forces acting on it that are not in balance then the **resultant force** will give rise to a change in velocity, i.e. an acceleration. **Newton's Second Law** states that $F = ma$, where F is the resultant force, m is the object's mass (kg) and a is the acceleration. The acceleration is always in the same direction as the resultant force. If several forces are acting then the **vector sum** of all the forces is the resultant force used in the equation, including friction and drag. **Newton's Third Law** states that 'if an object X exerts a force on object Y then object Y exerts an equal but opposite force on object X'. This, in effect, states that all forces appear in pairs. Note that the pairs of forces in Newton's Third Law always act on different bodies and are of the same type. This must not be confused with a **free body diagram** where all the forces are acting on the same body.

QUICK TEST

1. A city tram leaves a stop and travels 72 m in 12 seconds. What is the average (mean) speed of the tram during this time?

2. Determine the acceleration of the above tram assuming it accelerates uniformly.

3. A stone is dropped down a disused well and takes 3 seconds to reach the bottom. How deep is the well?

4. How fast is the stone travelling when it reaches the bottom of the well?

5. State a *suvat* equation that does not involve time.

6. A van's initial velocity along a straight road is $16\,\text{m}\,\text{s}^{-1}$ and its final velocity after 20 seconds is $26\,\text{m}\,\text{s}^{-1}$. What is its average (mean) velocity and how far has it travelled, assuming its acceleration was constant?

7. An object is dropped on Earth from a height of 2.45 m. Neglecting air resistance, how long does it take to hit the ground?

8. The same object is dropped from the same height above the Moon's surface ($g = 1.6\,\text{m}\,\text{s}^{-2}$). How long will the object take to reach the Moon's surface?

9. In questions **7** and **8** why is mass not important?

10. If an object shows a negative slope on a velocity–time graph, what does it indicate?

11. An object is moving at a constant velocity. How is this shown on a velocity–time graph?

12. Draw a displacement–time and velocity–time graph for an object thrown vertically upwards at a speed u and caught on the way down at the same height. Express the time to maximum height in terms of u.

13. A ball drops into soft sand with a velocity of $5\,\text{m}\,\text{s}^{-1}$. How far will it penetrate the sand if it decelerates at $125\,\text{m}\,\text{s}^{-2}$?

SUMMARY

- A distance–time graph shows how the speed of an object varies; a displacement–time graph shows how the velocity of an object varies

- The instantaneous speed or velocity of an object is the gradient of the line at that point in a distance–time or displacement–time graph (respectively); the direction of the velocity is given by the sign of the gradient

- The area under a velocity–time graph gives the displacement of an object

- A speed–time graph or a velocity–time graph shows how the acceleration of an object varies; the gradient of a line at any point provides the size and direction of the acceleration

- A negative acceleration is a **deceleration**

- An object in **free fall** is under the influence of a constant acceleration given by the **acceleration due to gravity of $9.81\,\mathrm{m\,s^{-2}}$. Air resistance will oppose motion until the weight equals the resistance, when the resultant force will be zero; the object is then said to have reached **terminal velocity**

- If the acceleration is constant or uniform, the four key **equations of motion, called the 'suvat'** equations, can be applied

- More complicated problems can be solved by splitting the problem into **two (or more) stages**

PRACTICE QUESTIONS

1. A cyclist accelerates away from a set of traffic lights uniformly to reach a velocity of $12\,\mathrm{m\,s^{-1}}$ in 9 seconds. They continue at this speed for 1.5 minutes before slowing down to a stop in 15 seconds.

   a) Draw a velocity–time graph for the cyclist's motion. **[3 marks]**

   b) Determine the cyclist's initial acceleration. **[2 marks]**

   c) How far did the cyclist travel in the first 9 seconds? **[2 marks]**

   d) What distance did the cyclist travel in total? **[2 marks]**

2. A model rocket of mass 100 g is launched vertically upwards with a thrust of 2.3 N.

   a) Calculate the acceleration of the rocket. **[3 marks]**

   b) The fuel was spent after 3.5 seconds. Assuming that the rocket climbed vertically with constant acceleration, determine (i) the speed of the rocket after 3.5 s and (ii) the height reached in this time. **[3 marks]**

   c) Calculate (i) the maximum height reached by the rocket and (ii) the time it takes the rocket to descend back to the ground. (Assume air resistance is negligible.) **[4 marks]**

3. A 747 passenger aircraft is powered by four engines each giving 220 kN of thrust. The mass of the aircraft at take-off is 320 tonnes.

   a) Calculate the total thrust of the aircraft. **[1 mark]**

   b) At take-off the air resistance is 50 kN. Determine the acceleration of the aircraft at take-off. **[3 marks]**

   c) The aircraft takes 15 minutes to reach its cruising speed of $250\,\mathrm{m\,s^{-1}}$. Calculate the average (mean) acceleration of the aircraft during this time. **[2 marks]**

   d) At cruising altitude the thrust is at 50% capacity and the aircraft is flying at a constant speed and at constant height. What is the lift force of the aircraft assuming its mass has now decreased to 300 tonnes and what is the resistive force or drag on the aircraft? **[3 marks]**

   e) The aircraft as it touches down to land is travelling at $45\,\mathrm{m\,s^{-1}}$ and decelerates at $2.0\,\mathrm{m\,s^{-2}}$. Determine (i) the time taken to come to a complete stop and (ii) the distance taken to come to a complete stop. **[4 marks]**

Projectile Motion

Velocity and Acceleration

A **projectile** is any object that has been dropped or has been projected and continues in motion under the force due to gravity. Projectiles may be launched vertically, horizontally or at an angle above the ground. Irrespective of the direction of launch, there are three principles that always apply to its motion:

● the components of motion in the **vertical** and **horizontal directions** are always independent of each other

● the **horizontal component** of the velocity is always constant (assuming that air resistance can be ignored) as the force due to gravity does not act in this direction and therefore has no effect on its motion

● as the force due to gravity acts in the vertical direction, the **vertical component** of the velocity is affected; hence the projectile's velocity in this direction is altered during flight.

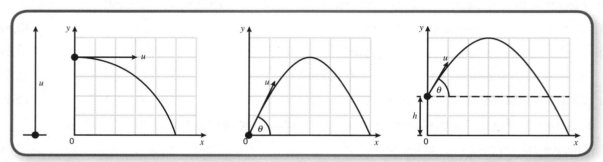

Vertical Projection

For a projectile launched vertically upwards (taken as the positive direction) the acceleration due to gravity, g is negative, and there is no horizontal component of the velocity. The 'suvat' equations of motion can be used to determine the velocity (v) and displacement (s) at any time (t) during its motion, giving

$$v = u - gt$$

$$s = ut - \frac{1}{2}gt^2$$

When a projectile reaches the top of its trajectory it becomes momentarily stationary, i.e. $v = 0$, and this concept is useful when determining its maximum height and time of flight.

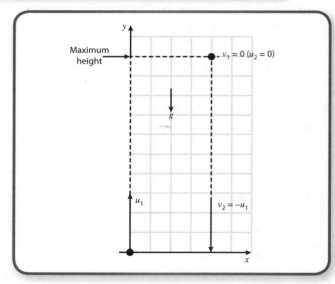

Horizontal Projection

When an object is projected horizontally above the ground its trajectory is a gradual arc that eventually takes it back to ground level; in fact, the path of the trajectory is always **parabolic**. Solving problems of this type requires the introduction of velocity components along the x and y directions, often written u_x and u_y for the initial velocities and v_x and v_y at time t later.

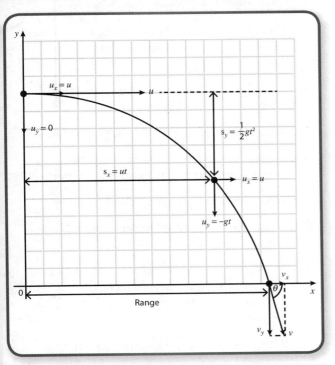

If an object is launched horizonally with a velocity u then at launch $u_x = u$ and $u_y = 0$. Its **horizontal velocity** is assumed to remain constant during its trajectory so after a time t it has travelled a distance (displacement) of $s_x = u_x t = ut$.

The component of the velocity in the vertical direction is now $v_y = -gt$.

The speed of the object at any time $v(t)$ can then be found via Pythagoras' Theorem as

$$v(t) = \sqrt{\left(v_x\right)^2 + \left(v_y\right)^2}$$

and the **direction** of the velocity (below the horizontal) is given by $\theta = \tan^{-1}\left(\frac{v_y}{v_x}\right)$. If T is the time of flight then the total horizontal distance travelled is called the **range**, given by $s = uT$. The time of flight is usually determined from the height of projection, h, since $h = \frac{1}{2}gT^2$. Note that the time of flight, T, is the only value that is the same for vertical and horizontal motion.

Projection at an Angle

When an object is projected at an angle θ above the horizontal then its path will always be **parabolic**. This is due to a constant horizontal motion that is subject

to a constant vertical acceleration acting downwards. If the initial velocity is u then the horizontal and vertical components of the initial velocity will be $u_x = u\cos\theta$ and $u_y = u\sin\theta$, respectively.

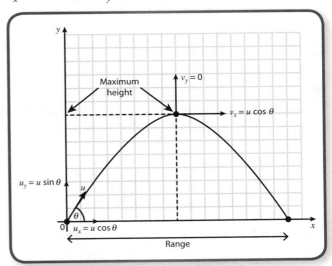

During the trajectory it is assumed that the horizontal component of the velocity remains the same, i.e. $v_x = u_x = u\cos\theta$, but the vertical component of the velocity is modified by the acceleration due to gravity, giving $v_y = u_y - gt$ or $v_y = u\sin\theta - gt$. The velocity of the object and its direction after a time t can then be found for all values of t again using Pythagoras' Theorem and $\theta = \tan^{-1}\left(\frac{v_y}{v_x}\right)$. The position or displacement of the object can then be derived using the 'suvat' equations: $s_x = v_x t = ut\cos\theta$ horizontally, and $s_y = u_y t - \frac{1}{2}gt$ or $s_y = ut\sin\theta - \frac{1}{2}gt^2$ vertically. Again, if T is the time of flight (given by $2u\sin\theta/g$) then the range of the projectile is $s_x = uT\cos\theta$ and the time of flight can be found by calculating the time to maximum height, where $v_y = 0$, using $v_y = u_y - gt$ and solving for t. The total flight time is double the answer, as the trajectory is symmetric. It is important to note that:

● For an object that is projected upwards from the ground, the velocity and angle when it returns to the ground will be identical (i.e. have the same magnitude), except that the vertical component of the velocity will be in a downward and not upward direction.

● For projections below the horizontal, a similar analysis can be carried out provided that the angle is assigned a negative value.

Drag forces increase with increasing speed; the effect on a projectile would be to reduce both its speed and range (displacement) as well as its maximum height. This results in a slightly asymmetrical path with a steeper descent. For most questions, however, drag can be ignored.

Other forces that may affect the motion, such as spin, are generally ignored.

For projectiles launched from a height above ground, such as a javelin, take the vertical height to be zero at point of projection. As the object lands below this height, the vertical component of displacement is taken to be negative.

SUMMARY

- **A projectile is any object that has been dropped or has been projected and continues in motion under the force due to gravity**

- **An object may be projected directly vertically, e.g. a tennis ball before being served, directly horizontally, e.g. a cannonball, from the ground at an angle to the horizontal, e.g. a golf ball, or from some height above ground, e.g. a javelin**

- **Drag is the name given to resistive forces experienced by objects travelling through air or a liquid**

- **Air resistance is a force that opposes motion when an object travels through the air; it is a type of drag**

- **Range is the maximum horizontal distance/displacement reached by a projectile**

- **The horizontal component of the velocity is considered to be constant throughout the trajectory**

- **The vertical component of the velocity is subject to the acceleration due to gravity, which acts vertically downwards**

- **Projectile motion follows a parabolic path**

- **A projectile at maximum height has a vertical velocity component of zero**

QUICK TEST

1. What is the initial vertical velocity component of an object that has been projected horizontally at $8\,\mathrm{m\,s^{-1}}$?

2. What is the size of the horizontal velocity component after 2 seconds?

3. Is the acceleration due to gravity constant?

4. Explain why it is possible to apply the equations of motion to projectile motion.

5. What forces act on a projectile?

6. What affects the vertical component of a projectile's motion?

7. If a projectile is launched vertically with a velocity of $12\,\mathrm{m\,s^{-1}}$, what is its velocity when it returns to the launch site? (Ignore air resistance.)

8. What is meant by the maximum height reached?

9. What is time of flight?

10. What angle is needed for a projectile to reach its greatest range?

11. An object is projected horizontally at a speed of $20\,\mathrm{m\,s^{-1}}$ from a sea cliff of height 30 m above sea level. (i) How long will it take to fall into the sea? (ii) What is the range of the object? (iii) Determine its speed as it hits the sea.

12. An arrow is fired at an initial speed of $24\,\mathrm{m\,s^{-1}}$ at an angle of 30° to the horizontal and at a height of 1.5 m above ground. Assume that it hits the target at the same height. (i) How long was it in flight? (ii) How far away is the target? (iii) What is the maximum height above ground?

PRACTICE QUESTIONS

1. A ball is thrown vertically upwards from a height of 2 m above the ground at a speed of 5 m s^{-1} and is caught at the same height on the way down. Ignore the effects of air resistance.

   a) Calculate the maximum height above the ground the ball reaches. **[3 marks]**

   b) Calculate the time of flight of the ball in the air. **[2 marks]**

   c) If air resistance cannot be ignored, what would be the effect on the maximum height reached? **[1 mark]**

   d) Calculate the speed of the ball if it is not caught and thus hits the ground. **[2 marks]**

2. A cannon on board an old ship is fired horizontally with a velocity of 110 m s^{-1} from the top deck, some 8 m above the water. Ignoring air resistance, calculate:

   a) the time taken to hit the water **[2 marks]**

   b) the range of the cannonball. **[1 mark]**

   c) If the cannonball hit a target some 30 m away, how high above the water line would it hit? **[3 marks]**

3. In an Olympic Games a hammer thrower launches the hammer ball at a velocity of 25 m s^{-1} at an angle of 38° to the ground from a height of 1.6 m.

   a) What assumption can be made about the hammer and its trajectory through the air? **[1 mark]**

   b) Determine the horizontal and vertical components of the initial velocity. **[2 marks]**

   c) For how long is the hammer in the air? **[6 marks]**

   d) Calculate how far the hammer will travel. **[1 mark]**

   e) What angle would be needed in order for the hammer to have a greater range? **[1 mark]**

Energy, Work and Power

Types of Energy

A considerable amount of physics involves objects that possess energy. The principal types of energy are shown in the table. Two key types are the energy concerned with moving objects, called **kinetic energy**, and the energy concerned with the position of an object in a gravitational field, called **gravitational potential energy**. Kinetic energy is given by $E_k = \frac{1}{2}mv^2$, where m is the object's mass and v is its velocity. Potential energy (in this example, the gravitational potential energy) is $E_p = mgh$, where g is the acceleration due to gravity and h is the height gained above a fixed point. All energies are **scalar** quantities and are measured in joules (J). One joule is the energy needed to raise a 1 N weight through a vertical height of 1 m.

Energy	Applies to an object
Gravitational potential energy	Position
Kinetic energy	Movement
Thermal energy	Heat
Elastic energy	Compressed or stretched state

Conservation of Energy

One of the key conservation laws in physics is the **conservation of energy**. This law states that energy cannot be created or destroyed, but can only be transformed or transferred from one form into another. For example, electrical energy is transformed into thermal energy when a kettle full of water is boiled.

The **total amount of energy**, e.g. kinetic energy and potential energy, remains **constant** in a closed system.

$$E_k + E_p = \text{constant}$$

An object projected upwards loses kinetic energy but gains potential energy; at the top of its trajectory the object will have transformed all its kinetic energy into gravitational potential energy, and by the time it has fallen back down again it will have transformed all its gravitational potential energy back to kinetic energy.

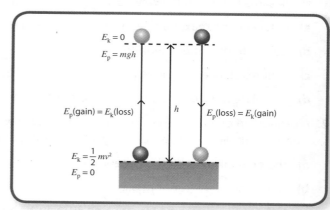

If no energy escapes during the transfer process, e.g. as sound, heat or light energy, then the system is known as a **closed system**.

Energy Efficiency

In a closed system the transfer of energy is 100% efficient as no energy is lost to the outside. However, in the 'real' world energy is lost in the transfer process, perhaps through heat, or sound or light. In such processes the ratio of the energy output to the energy input is known as **efficiency**. The efficiency, η, as a percentage is given by

$$\eta = \frac{\text{output energy}}{\text{input energy}} \times 100\% \text{ or } \eta = \frac{\text{work done}}{\text{energy supplied}} \times 100$$

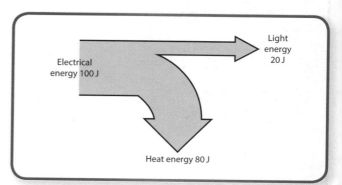

It is often convenient to represent the energy changes and transfers in a **Sankey diagram**. This is a flow diagram that indicates the changes in energy by showing arrows of widths that are directly proportional to the amounts of energy involved.

Work Done

In physics questions it is assumed that processes involving energy transfer take place in a closed system, unless it is specifically mentioned in the question that energy is lost. Under these conditions, in a mechanical system the **energy transfer** is known as **work done**, W. The work done depends on the force, F, and the distance, s, that the object is moved by the force in the same direction as the force, and is given by

$$W = F \times s = Fs$$

where W is in joules, F in newtons and s in metres. This means that the units of work done, i.e. N m, is the same as joules. When the force moving the object is not in the same direction that the object moves, the component of the force in that direction needs to be used, i.e.

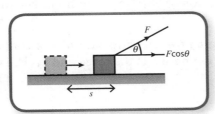

$$W = F \cos\theta \times s = Fs \cos\theta$$

Force–Distance Graphs

If the force is constant and a graph is drawn of the force against the distance moved then the area under the **force–distance graph** represents the amount of **work done**. If a variable force acts on an object and causes it to move in the direction of the force then the area under the force–distance graph represents the total work done. It is often necessary to add up the number of squares under the curve to determine the total area. For example, the force used to stretch a spring elastically gives rise to Hooke's Law. The area under the graph of force against extension is then the work done in stretching the spring and gives the **elastic potential energy** stored in the spring.

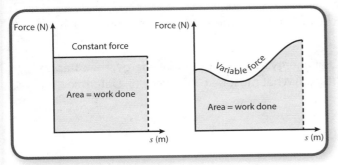

Power

Power is the amount of work done in a specific amount of time; it is also the amount of energy that has been transferred in that time. It is commonly referred to as the **rate of doing work**. Power has units of **watts**, where $1\,W = 1\,J\,s^{-1}$.

$$\text{power } (P) = \frac{\text{work done } (W)}{\text{time taken } (t)} = \frac{\Delta W}{\Delta t} \quad \text{or} \quad P = \frac{\Delta E}{\Delta t}$$

Using the fact above that work done is force $\times$ distance moved, the following useful result is obtained:

$$\text{power } (P) = \frac{Fs}{t} = F \times \frac{s}{t} = F \times v \quad \text{or} \quad P = Fv$$

i.e. **power** is the **force (in the direction of motion)** $\times$ **speed**. There are many types of power such as muscle power, electrical power, solar power, hydroelectric power and engine power.

SUMMARY

- The **kinetic energy** of a moving object of mass m and speed v is given by $\frac{1}{2}mv^2$
- The change in **gravitational potential energy** of an object of mass m near the Earth's surface that is moved through a vertical distance h is given by mgh
- Kinetic and potential energy are **scalar** quantities and measured in **joules**
- **Conservation of energy** states that energy cannot be created or destroyed but can only be converted or transferred from one form into another
- **Work done** is achieved when a force moves an object a distance s in the direction of the force
- The **work done** by a force is calculated by the area under the **force–distance graph**
- **Efficiency** is the ratio of the output energy from a device to the energy input and it is often expressed as a percentage; a pictorial representation of efficiency can be shown in a **Sankey diagram**
- **Power** is the rate of doing work or the rate of energy transfer and has units of watts $\left(1\,\text{W} = 1\,\text{Js}^{-1}\right)$; for a moving object it can be expressed by $P = Fv$, where F is the force and v is the speed of the object

QUICK TEST

1. A student of weight 600 N climbs a flight of stairs to a height of 15 m. What is the gain in potential energy?

2. A car of mass 1000 kg is moving at $25\,\text{m s}^{-1}$. How much kinetic energy does it have?

3. A ball of mass 0.40 kg is thrown vertically upwards at a speed of $5\,\text{m s}^{-1}$. Calculate: (i) the initial kinetic energy (ii) the maximum gain in potential energy (iii) the height the ball reaches.

4. Calculate the amount of work done when a weight of 700 N ascends a height of 3 m.

5. The engines of an aeroplane produce a thrust of 48 kN while in level flight maintaining a speed of $100\,\text{m s}^{-1}$. What is the output power of the engines?

6. A small filament bulb is supplied with 25 J of energy but only 3 J is emitted in the form of light energy. (i) What is the efficiency of the filament bulb? (ii) What other main source of energy loss accounts for this inefficiency?

7. A power station is only 35% efficient. It provides 58 MW of power to the electricity grid. How much energy is produced by the power station?

8. A catapult's elastic band is stretched and stores 0.5 J of energy as elastic potential energy; the catapult then launches a small wooden sphere of mass 0.008 kg. (i) What must be conserved? (ii) Determine the speed of the wooden sphere.

PRACTICE QUESTIONS

1. A hydroelectric power station uses water from an upper lake which falls through a height of 65 m.

   a) Calculate the change in potential energy for a mass of 1.0 kg of water falling through this vertical height. **[2 marks]**

   b) Determine the maximum power available if water flows through the power station at a rate of 4.4×10^7 kg per hour. **[3 marks]**

   c) Only 5.5 MW is provided to the electricity grid. Calculate the overall efficiency of the power station. **[2 marks]**

2. A space diver of mass 70 kg jumps from a balloon at an altitude of 41.4 km and is in free fall for 37.6 km.

   a) Calculate the potential energy lost during the free fall (assume no air resistance). **[2 marks]**

   b) Determine the maximum possible speed achieved during the free fall stage (give your answer in $km\,h^{-1}$). **[3 marks]**

   c) The maximum speed achieved is actually $1300\,km\,h^{-1}$. Give reasons why this speed is substantially less than that calculated in part **b)**. **[2 marks]**

   d) This difference in maximum kinetic energy is equal to the work done against air resistance. Calculate the work done against resistance. **[2 marks]**

   e) Determine the mean force due to air resistance acting on the space diver. **[2 marks]**

3. An aircraft of mass 120 000 kg is travelling at a constant speed of $220\,m\,s^{-1}$ at a constant height.

   a) Determine the up-thrust force (lift) necessary to keep the aircraft at a constant height. **[2 marks]**

   b) The drag on the aircraft at this height is 80 kN. Calculate the power required in each of its four engines to maintain its constant speed. **[3 marks]**

   c) If the pilot reduced the power to each engine, what would be the effect on the aircraft? **[2 marks]**

Momentum and Collisions

Linear Momentum

The **linear momentum** of an object (in motion in one dimension) is its **mass × velocity**. It is a vector with units of $kg\,m\,s^{-1}$ or $N\,s$. The symbol for linear momentum is p, so the equation is $p = mv$. Momentum is an important concept in physics and is involved in all types of collisions, from the smallest particles to the largest galaxies. A small mass travelling very fast may have the same momentum as a large mass travelling very slowly.

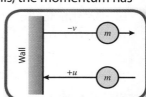

Newton's Second Law and Momentum

Newton's Second Law states that $F = ma$.
Substituting $a = \frac{v-u}{\Delta t}$ into the Second Law gives

$$F = \frac{mv - mu}{\Delta t} = \frac{\Delta(mv)}{\Delta t} = \frac{\Delta p}{\Delta t}$$

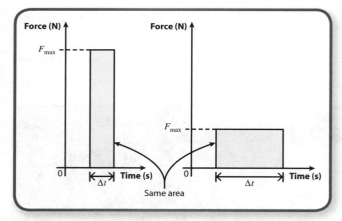

Same area

This allows Newton's Second Law to be restated in terms of momentum. The above equation states that the force of impact between two colliding objects is equal to the change of momentum that occurs during unit contact time, i.e. the impact force is equal to the **rate of change of momentum**. The shorter the contact time, the larger the impact force. Conversely, the longer the contact time, the lower the impact force, and this is used in the design of cars that uses crumple zones, airbags and seatbelts to reduce the impact forces involved in any collisions or by rock climbers using nylon ropes that stretch. Rewriting this equation gives $F\Delta t = \Delta(mv) = \Delta p = m\Delta v$.

The **impact force × contact time** defines the **impulse**, I, of a force acting on an object and is equal to the change of momentum of the object.

When objects rebound from walls head on, the velocity and thus the momentum change sign. The impact force is then $F = \frac{-mv - mu}{\Delta t}$ and, if there is no loss of speed, this reduces further to $F = -\frac{2mu}{\Delta t}$.
For oblique collisions with walls, the momentum has to be resolved into two components parallel and perpendicular to the wall. In all these impacts it will be assumed that the mass remains constant.

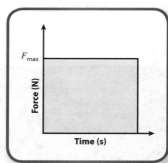

Force–Time Graphs

If a constant force acts on an object for a time t then a graph of F against t gives a straight-line graph. The **area under the graph** then represents the change of momentum. In this context, it is common to use the units $N\,s$. For real impacts there are substantial variations in the impact force with time. Evaluating the area under the graph is more complex but the standard method of counting squares provides a reasonable result.

Conservation of Momentum

When two (or more) particles collide, not only is energy conserved but also momentum. The **principle of conservation of momentum** states that the total momentum before a collision is the same as the total momentum after the collision. Another definition states that the total momentum remains constant provided that no external forces act on the system. Mathematically, the conservation of momentum is given as

total initial momentum = total final momentum
$$m_1 u_1 + m_2 u_2 = m_1 v_1 + m_2 v_2$$

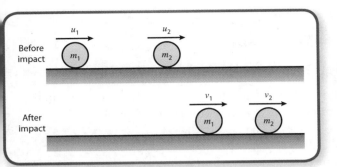

Before impact: u_1, m_1; u_2, m_2

After impact: v_1, m_1; v_2, m_2

Because momentum is a vector it is important to ensure that the correct signs are used when using the equation. Note also that in some collisions the two particles stick together (or **coalesce**). In this special case they will both move off with a velocity V so that the equation becomes

$$m_1 u_1 + m_2 u_2 = (m_1 + m_2)V$$

Conservation of momentum is a consequence of Newton's Third Law, which discusses the interaction between two objects resulting in equal but opposite forces acting on each other.

Elastic and Inelastic Collisions

There are three types of collisions that lead to a different amount of energy transfer; these collisions are classified as being **elastic**, **partially inelastic** or **totally inelastic** and the classification depends solely on the amount of **kinetic energy lost**.

Elastic	Partially inelastic	Totally inelastic
No loss of kinetic energy	Partial or total loss of kinetic energy	
Objects move apart	Objects move apart	Objects stick together

To determine whether a collision is elastic or inelastic, the kinetic energy of each object before and after the collision must be worked out. For example, if two objects collide, the kinetic energy is worked out for each particle before the collision and again after the collision using $E_k = \frac{1}{2}mv^2$. The total kinetic energy both before and after the collision is calculated and this determines the nature of the collision, i.e.

$$\frac{1}{2}m_1 u_1^2 + \frac{1}{2}m_2 u_2^2 = \frac{1}{2}m_1 v_1^2 + \frac{1}{2}m_2 v_2^2$$

Another example is an object of mass m that is dropped from a height H and rebounds to a height h. From energy considerations, there are two scenarios:

● the loss of gravitational potential energy = kinetic energy immediately before impact, i.e. $mgH = \frac{1}{2}mu^2$ or $gH = \frac{1}{2}u^2$

● the gain in gravitational potential energy on rebounding to height h = loss of kinetic energy after the impact, i.e. $mgh = \frac{1}{2}mv^2$ or $gh = \frac{1}{2}v^2$.

So the ratio $\frac{h}{H}$ gives $\left(\frac{v}{u}\right)^2$, i.e. the fraction of initial kinetic energy that is recovered as kinetic energy after rebounding from the ground. If $h = H$ then the ratio is 1 and there is no loss of kinetic energy, i.e. the rebound is elastic.

Changes in the energy in a collision are particularly important in explosive events.

SUMMARY

● **Momentum** is the product of mass and velocity, $p = mv$; units can be expressed in $kg\,m\,s^{-1}$ or $N\,s$

● **Linear momentum** is the momentum of an object that moves only in one dimension

● **Conservation of momentum** states that the total momentum of a colliding system or explosion remains constant

● An **elastic collision** means that no kinetic energy is lost; an **inelastic collision** means that some kinetic energy is lost and transferred into heat, light and other forms of energy

● **Impulse** is the product of the applied force and the contact time, i.e. $I = F\Delta t$, and is equivalent to the change in momentum: $I = \Delta p = mv - mu = m\Delta v$

● The area under a **force–time graph** gives the change in momentum

1. A snooker ball of mass 0.1 kg is moving at a speed of $5\,m\,s^{-1}$. Calculate the momentum of the ball.

2. The above snooker ball collides directly with another ball that is stationary; the first ball moves off in the same direction with a speed of $3.5\,m\,s^{-1}$. Calculate the change in momentum of the first ball.

3. State Newton's Second Law in terms of momentum.

4. A ball of mass 0.15 kg is moving at $35\,m\,s^{-1}$ when it hits a wall head on and rebounds at a speed of $27\,m\,s^{-1}$. Calculate the change in momentum.

5. If in the above question the ball was in contact with the wall for 12 ms, (i) give the equation for the impact force and (ii) calculate this impact force.

6. A molecule of mass 4.2×10^{-26} kg collides head on with a flat surface at a speed of $380\,m\,s^{-1}$ and rebounds in the opposite direction. The contact time is 0.15 ns. Calculate (i) the change in momentum and (ii) the force on the molecule (assume the collision is elastic).

7. A railway engine of mass 6.0 tonnes is moving at $4.5\,m\,s^{-1}$ when it collides with a stationary empty wagon of mass 1.2 tonnes. Calculate the speed of both trains if they couple together on impact.

8. Give a relevant example of a safety feature on a car and give the reasons for it being a safety feature.

9. A shell of mass 2.45 kg is fired from a stationary tank with a speed of $125\,m\,s^{-1}$. The mass of the tank is 4500 kg. Calculate the recoil of the tank when the shell is fired.

10. What does the area under a force–time graph represent? Give the units.

PRACTICE QUESTIONS

1. The engine of a goods train has a mass of 6.5×10^4 kg and is moving at a slow speed of $0.4\,\text{m s}^{-1}$ in order to couple to its load of four wagons, which are all stationary. The total mass of the load wagons is 8.5×10^4 kg.

   a) Calculate the momentum of the engine. [2 marks]

   b) The engine and wagons couple together and then move together. Calculate the velocity of the engine and wagons after the coupling. [3 marks]

2. a) State two quantities that are conserved in elastic collisions. [2 marks]

   b) Two snooker balls collide head on along the same straight line. The white ball is travelling at $5\,\text{m s}^{-1}$ and hits a red ball that is stationary. Both balls have a mass of 100 g.

   (i) Calculate the initial momentum of the white snooker ball. [1 mark]

   (ii) Determine the speed of the red ball after the collision if the white ball moves off at a speed of $1.5\,\text{m s}^{-1}$ along the same straight line. [3 marks]

   c) The red snooker ball hits the cushion head on and rebounds from the cushion without loss of speed in 0.07 s. Determine the impact force on the cushion by the ball. [2 marks]

3. A cricketer uses his bat to hit a ball travelling towards him at $40\,\text{m s}^{-1}$. The mass of the ball is 0.16 kg and the force–time graph of the collision is shown.

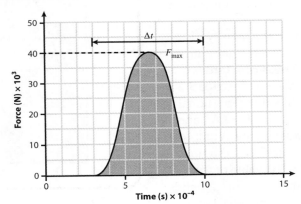

   a) From the graph, determine the average (mean) impact force. [1 mark]

   b) Calculate the impulse between bat and ball. [2 marks]

   c) Using this information, determine the speed of the ball if it returns in the opposite direction and along the same straight line. [4 marks]

4. a) What is meant by an inelastic collision? [1 mark]

   A particle of mass 1.5 kg moving with a speed of $6.5\,\text{m s}^{-1}$ collides head on with another particle of mass 0.5 kg travelling at $4.0\,\text{m s}^{-1}$. The smaller particle moves away with a speed of $3.5\,\text{m s}^{-1}$.

   b) Using conservation of momentum, determine the velocity of the larger particle after the collision. [3 marks]

   c) How much kinetic energy is lost in the collision, and what happens to this energy? [4 marks]

Bulk Properties of Materials

Force–Extension Graphs and Hooke's Law

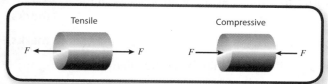

Macroscopic or **bulk** properties of materials include density, elasticity, strength and stiffness. Solid materials can be broadly classified into four groups: crystalline solids, amorphous solids, glasses and polymers. Testing these materials by pulling them apart with **tensile forces** or pushing them together with **compressive forces** reveals the nature of their bulk properties. The simplest bulk property is **density**, which is defined as the mass per unit volume (units of $kg\,m^{-3}$), i.e. $\rho = \frac{m}{V}$. The density of water is $1000\,kg\,m^{-3}$ and **relative density** is the density of a material relative to the density of water. Another bulk property of a solid is its **stiffness constant** (k). For most materials that are stretched under a **load** (F), the force applied is initially directly proportional to the **extension** (Δl). This is written as $F = k\Delta l$, and the equation is referred to as **Hooke's Law**. A graph of load against extension therefore produces a straight line that goes through the origin. The value of the gradient gives the stiffness constant and this depends on the microscopic properties of the material being stretched; it has units of $N\,m^{-1}$.

Behaviour of Ductile Materials

A **graph** of F against Δl for different materials reveals very different macroscopic behaviour; all show Hooke's Law type behaviour up to a point and this includes metals, fibres, rubbers and polymers. A typical force–extension graph for a metal wire shows linear behaviour up to a point called the **limit of proportionality**, before deviating beyond a point called the **elastic limit**, where **permanent deformation** occurs. In this **plastic region**, a material may exhibit what is called **ductile behaviour**, i.e. the ability to retain strength while changing shape, usually into a very thin wire.

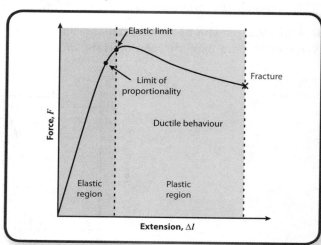

Behaviour of Brittle Materials

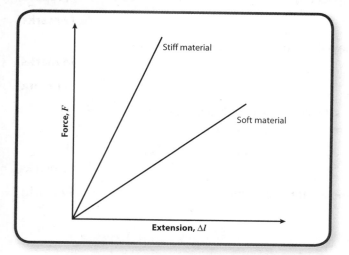

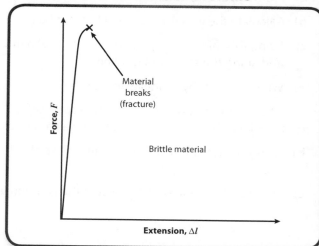

In other materials, beyond the elastic limit the material breaks, indicating a more **brittle** type behaviour. Materials that are brittle include glass and ceramics, as well as bricks and pottery. Generally, brittle materials have much steeper gradients in their force–extension graphs, indicating that they are a lot stiffer than metals. Even in ductile materials there is a point when the material finally **breaks** or **fractures**. The nature of the fracture is very different between ductile and brittle materials.

Modelling Bonds between Atoms

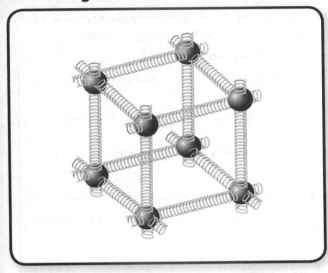

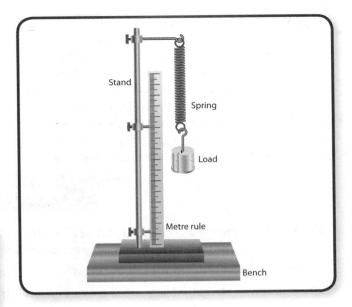

Metallic wire (showing ductile behaviour) obeys Hooke's Law because the **bonds** that hold the atoms together act like **springs**. The way springs, or a combination of springs, behave gives an insight into the behaviour of ductile materials. A **spring constant** is a measure of how hard it is to stretch a spring. A large spring constant means that the spring is **stiff**, and the spring constant has the same units as the stiffness constant (opposite). The spring constant can be measured by a simple experiment where a spring is stretched and its extension is recorded as the load is increased (as shown). A graph of force (load) against extension reveals both elastic behaviour as well as plastic behaviour beyond the elastic limit.

Springs in Parallel and Series

Combining springs gives a much better insight into how atoms behave. Springs can be combined by joining two springs together either in a **series** or in a **parallel** arrangement. Under these conditions, there is an **effective spring constant**, k_{eff}, that is related to the total extension of the spring arrangement. For two springs with spring constants k_1 and k_2, if they are arranged in series the equation is $\frac{1}{k_{eff}} = \frac{1}{k_1} + \frac{1}{k_2}$, and if they are arranged in parallel it is $k_{eff} = k_1 + k_2$.

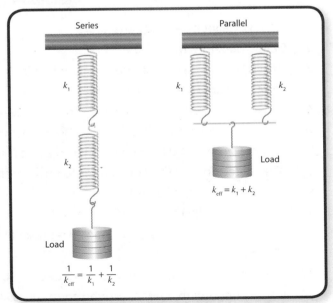

Beyond the Elastic Limit

A wire will initially obey Hooke's Law because the bonds between the metal atoms act like **springs**. When a force or load is applied to the wire, the wire **stretches** as the bonds extend. Taking the force away allows the wire to return to its **original** length and so do the bonds. Beyond the **elastic limit**, however, the forces involved are enough to break some of the bonds holding the atoms together and this allows the atoms to begin to slip past each other, assisted by the movement of structural defects called **dislocations**. This is the plastic region and the process of atoms sliding past each other gives rise to the property known as **ductility**. This process allows many metals to be formed into very thin wires. At some point, the material deforms so much that it will form cracks that spread and grow, eventually leading to **fracture**.

SUMMARY

- **Hooke's Law states that the force applied to a wire or spring (the load) is directly proportional to the extension; the constant of proportionality is the** stiffness constant **or** spring constant
- **Hooke's Law only applies in the** elastic region **up to the** limit of proportionality
- **A material stretched beyond the** elastic limit **it is permanently deformed; this region is known as the** plastic region
- **Materials can be described as either** ductile **(permanent change of shape without breaking) or** brittle **(shows no plastic region)**
- **The** limit of proportionality **is the end point of the linear region of a force–extension graph**
- **The units of the spring or stiffness constant are** $N\,m^{-1}$
- **The bonds that hold materials together can be modelled as** springs **between adjacent atoms**
- **Springs can be combined in** series **or** parallel **arrangements to give an effective spring constant**

QUICK TEST

1. What is meant by elastic behaviour?
2. Define Hooke's Law.
3. How is Hooke's Law displayed on a graph of force against extension?
4. What is meant by a ductile material?
5. What does the gradient correspond to in a force–extension graph?
6. What is a good representation of the bonding forces between atoms?
7. How are stronger bonds represented on a force–extension graph?
8. What defect in a metal allows atoms to move past each other?
9. What are the units of density?
10. What is meant by the term 'tensile force'?
11. What does the limit of proportionality represent?
12. What is the elastic limit?
13. If a spring extends by 10 cm under a force of 72 N, what is the value of the spring constant?
14. If two springs (with the same spring constant as above) are joined together in parallel, what is the effective spring constant of the two springs?

PRACTICE QUESTIONS

1. a) What is meant by the density of a material? [1 mark]

The metal alloy called brass is an alloy of copper and zinc (70% copper and 30% zinc by volume). The density of copper is $8.9 \times 10^3 \, kgm^{-3}$ and the density of zinc is $7.1 \times 10^3 \, kgm^{-3}$.

b) Calculate the mass of copper and the mass of zinc needed to make brass that has a volume of $2.2 \times 10^{-5} \, m^3$. [3 marks]

c) Use these values determine the density of brass. [2 marks]

2. a) State Hooke's Law. [2 marks]

The diagram shows the force–extension graph of two materials, A and B.

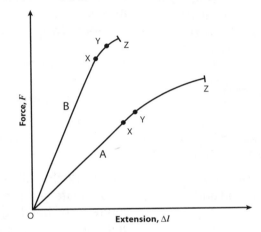

b) What do the positions marked X, Y and Z correspond to? [2 marks]

c) State the physical property represented by the gradient of the section OX of the graph. [1 mark]

d) Which of the two materials displays the property known as ductility and what is this region called? [2 marks]

e) Describe the properties of the other material and give an example of this material. [4 marks]

3. A piece of exercise equipment consists of a set of parallel and identical springs. One spring is stretched and a force of 8 N is needed to stretch the spring by 5 cm.

a) Calculate the spring constant for this single spring. [2 marks]

b) Determine the force required to stretch a single spring by 25 cm. [1 mark]

c) Calculate the amount of stretch that three parallel springs would produce if a force of 100 N is applied. [3 marks]

d) What assumption has been made in arriving at these answers? [1 mark]

Stress, Strain and the Young Modulus

Tensile Stress and Strain

Force–extension graphs depend on the dimensions of the material. A more appropriate graph that eliminates the dimensions of the material is obtained by plotting a stress–strain curve. **Tensile stress** is a measurement of the **force per unit area** (the cross-sectional area). The force (as in the case of Hooke's Law) is the tension exerted on the wire. Tensile stress is given by

$$\text{tensile stress} = \frac{\text{force}}{\text{cross-sectional area}} = \frac{F}{A}$$

with units of $N\,m^{-2}$ or pascals, Pa.

Stress is often given the symbol σ, where $\sigma = \frac{F}{A}$, i.e. the force or tension per unit cross-sectional area.

Tensile strain is the ratio of the extension to the original length of the material under a load and is given by

$$\text{tensile strain} = \frac{\text{extension}}{\text{original length}} = \frac{\Delta l}{l}$$

Strain is often given the symbol ε, where $\varepsilon = \frac{\Delta l}{l}$. Strain is extension per unit length; it has no units but it is sometimes given as a percentage, so that a 10% strain means $\frac{\Delta l}{l} = 0.1$.

The Young Modulus

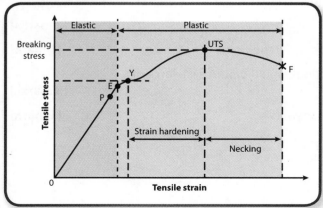

Plotting tensile stress on the y-axis and tensile strain along the x-axis will show how the stress on a wire varies with strain. From 0 to the **limit of proportionality** (P), the stress is directly proportional to the strain. This constant value (the gradient of the line) is comparable with the stiffness constant in Hooke's Law but is now independent of the dimensions of the sample. It is called the **Young modulus**, E, of the material.

$$\text{Young modulus, } E = \frac{\text{tensile stress }(\sigma)}{\text{tensile strain }(\varepsilon)}$$

$$\text{and} \quad E = \frac{(F/A)}{(\Delta l/l)} = \frac{Fl}{A\Delta l}$$

Comparing the values of the Young modulus in various materials gives valuable insight into the nature of the bonding. As strain has no units, the units for the Young modulus are $N\,m^{-2}$ or Pa. The values tend to be very large, so MPa and even GPa are commonly used.

Information from Stress–Strain Graphs

A **stress–strain curve** provides detailed information about the properties of a material; it looks very similar to a force (load)–extension graph. For example, a metal such as copper continues beyond P to show a slight curve that extends to the **elastic limit** (E). This is the point where, if the tension is removed, the wire would still return to its original length. Beyond E, the wire is permanently stretched and shows **plastic deformation**; the stress–strain curve becomes more distinctive up to and beyond the **yield point** (Y), where the wire begins to show weakness. After Y, a small change in stress causes a rapid change in strain and this indicates **plastic flow**. The **maximum stress** or **strength** of a material is given by the position of the **ultimate tensile stress** (UTS); this point is also sometimes called the **breaking stress**. Beyond UTS, the wire becomes narrower at its weakest point and the reduced cross-sectional area leads to a process called '**necking**' that ultimately leads to **fracture** (F).

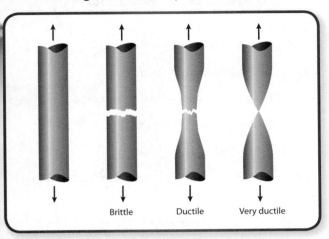

Brittle   Ductile   Very ductile

Measuring the Young Modulus

Measurements for stress–strain curves can be made using an experimental set-up similar to that shown. Long lengths of wire, typically 2–3 m long, are used and the extension measured using Vernier scales. A micrometer is used to determine the diameter of the wire. When measuring the Young modulus, it is common for a material in this elastic region to continue to stretch over a period of time when under a constant stress, a phenomenon known as **creep**.

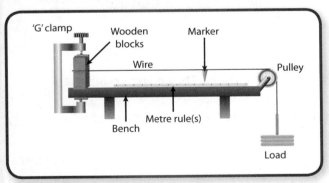

'G' clamp   Wooden blocks   Marker
Wire   Pulley
Metre rule(s)
Bench   Load

Stress–Strain Curves

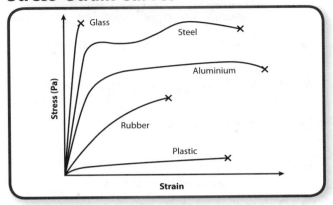

Stress–strain curves provide a method of describing the essential properties of materials. Typical stress–strain curves for **glass** (brittle), **steel** and **aluminium** (ductile metals), and **rubber** and **plastic** (polymers) are shown. Plastics tend to have a very short elastic region but an exceptionally long plastic region. Missing from the diagram are ceramics. **Ceramic** materials, such as pottery and bricks, are very brittle (but extremely strong) materials that tend to crumble when high stress causes a small strain. As such, they possess large values of the Young modulus and their stress–strain gradients are even steeper than that of glass in their elastic regions. Materials that can withstand high stress and strain before fracturing are described as tough. The table shows the values of the Young modulus for a range of materials.

Material	Young modulus
Metals	120 GPa
Glass	70 GPa
Pottery	360 GPa
Rubber	0.1 GPa
PVC	4 GPa
High-carbon steels	200 GPa

SUMMARY

- Tensile stress is $\frac{\text{force}}{\text{cross-sectional area}}$, with units of $N\,m^{-2}$ or Pa

- Tensile strain is $\frac{\text{extension}}{\text{original length}}$; strain has no units

- The Young modulus is $\frac{\text{tensile stress}}{\text{tensile strain}}$ (units of Pa)

- Properties of materials can be predicted from their stress–strain curves

- The gradient of the linear part of a stress–strain curve gives the Young modulus

- Ultimate tensile stress (also called the breaking stress) is the maximum stress and is a measure of the strength of a material

- Creep occurs in ductile materials when it continues to stretch under a constant stress

- Necking occurs prior to fracture and often at a lower stress value than the UTS owing to a reduced cross-sectional area

- Different materials have a wide range of values of the Young modulus, typically between 0.1 GPa and 400 GPa

QUICK TEST

1. What is the difference between tensile stress and compressive stress?

2. How is tensile strain defined?

3. What are the units of the Young modulus?

4. Define the Young modulus.

5. What is meant by ultimate tensile stress?

6. What process occurs before a ductile material fractures?

7. How is the strength of a material determined?

8. If one material is stiffer than another, how would this be shown on a stress–strain graph?

9. What is meant by creep?

10. Which is stronger, glass or ceramic? Give a brief reason.

11. Determine the Young modulus for steel if the tensile stress is 7.8×10^8 Pa and the tensile strain is 3.7×10^{-3}.

12. What kind of stress–strain curve would be obtained for a rubber material?

PRACTICE QUESTIONS

1. The stress–strain curves for two materials, A and B, are shown in the diagram. The end points represent fracture.

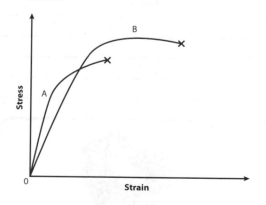

a) State and explain, giving the reasons for your choice, which material is:

   (i) stiffer [2 marks]

   (ii) tougher [2 marks]

   (iii) stronger [2 marks]

   (iv) more ductile. [2 marks]

b) On the graph, sketch a third material that is brittle, stronger than both A and B, and has a lower Young modulus than both A and B. Label your curve as C. [3 marks]

2. The stress–strain curve for a metallic alloy of brass is shown, and three regions are labelled A, B and C.

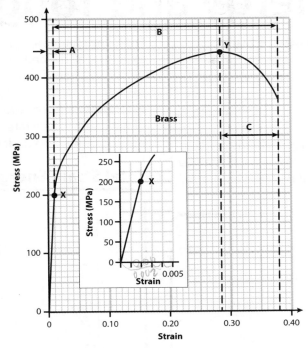

a) Using the letters from the graph, which region shows:

   (i) plastic flow

   (ii) necking [3 marks]

   (iii) elastic deformation?

b) What do the points labelled X and Y represent? [2 marks]

c) Using the enlarged part of the graph, determine the Young modulus of the brass alloy, giving your answer in MPa. [3 marks]

3. A coil of steel wire is being used to fence off an area of land. The steel wire has a diameter of 3.0 mm.

   a) Calculate the cross-sectional area of the steel wire. [2 marks]

   b) Between each wooden post, the steel wire is subject to a force of 1750 N to tension each of the wires. Calculate the tensile stress in the wire. [2 marks]

   c) For one side of the land, a piece of steel wire of length 45 m is used. Determine the extension of the wire when it is under tension. The Young modulus of steel is 210 GPa. [3 marks]

Strain Energy and Toughness

Strain Energy and Hooke's Law

When a material is stretched by opposing forces (tensile forces), the material is said to be **strained**, and work has to be done by external forces to cause this strain. If the elastic limit is not exceeded, removing the force (tension) allows the object to do work to return to its original shape. This means that energy is stored in the material as potential energy when it is strained and this is called **elastic potential energy** or simply the **strain energy**.

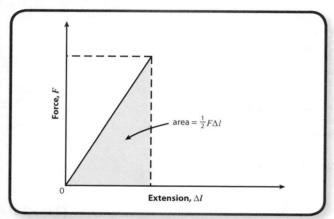

In the case of a spring, the energy stored in a spring is the **work done** in stretching the spring. The work done depends on the force that has been applied and the amount of extension produced. If a force is suddenly removed from a spring, the elastic energy stored in it is converted to kinetic energy of the spring. In a **force–extension graph**, the work done is the area under the straight line, i.e. the area of the triangle,

$$\text{work done} = \frac{1}{2} \times \text{force} \times \text{extension} = \frac{1}{2}F\Delta l$$

and this is the **strain energy**. Provided that F is in newtons and Δl is in metres, the strain energy is in joules. The strain energy can also be expressed in terms of the spring constant. Using Hooke's Law $F = k\Delta l$ and substituting for F gives

$$\text{elastic strain energy} = \frac{1}{2}k(\Delta l)^2$$

For many materials, the force–extension graph extends well beyond the elastic limit. However, work is still being done to produce an extension and the strain energy stored in the material is the **total area under a force–extension graph**. The nature of the graph will dictate what approach is taken to determine the area under the curve.

Total Strain Energy

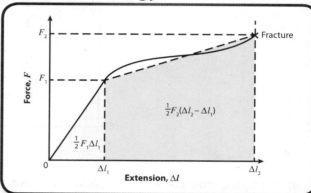

In the case shown, the material obeys Hooke's Law up to a force of F_1 and extension Δl_1, followed by plastic behaviour up to its fracture limit of F_2 with an extension Δl_2. The **total work done** is the area under the graph that obeys Hooke's Law plus the second area where Hooke's Law is no longer obeyed. In this region, energy is still required to stretch the material and therefore work is still being done. It is quite usual under these circumstances to approximate the shape to allow for an easier calculation, or to count squares. In this example, the elastic strain energy is given by the area of the triangle and the area of the trapezium:

$$\text{elastic strain energy} = \frac{1}{2}F_1\Delta l_1 + \frac{1}{2}F_2(\Delta l_2 - \Delta l_1)$$

Toughness and Strain Energy Density

The degree of **toughness** of a material is a measure of the energy needed to break or fracture it. In the example above, the toughness is given by the total strain energy up to the point of fracture. Toughness is expressed in joules. Another important parameter, called the **strain energy density**, gives the strain energy per unit volume of material. This is a measure of the amount of energy stored in a material that does not depend on its dimensions. This is obtained directly from the **stress–strain curve**. If the original length of wire is l and the cross-sectional area is A then the volume of wire is Al. The strain energy density is then given by

$$\text{strain energy density} = \frac{\frac{1}{2}F\Delta l}{Al}$$

$$= \frac{1}{2}\left(\frac{F}{A}\right)\left(\frac{\Delta l}{l}\right) = \frac{1}{2}\text{stress} \times \text{strain}$$

i.e. the area under a stress–strain curve. Strain energy density has units of $J\,m^{-3}$, which is the same as pascals.

Stress–Strain Graphs: Loading and Unloading

When materials are **loaded** and then **unloaded**, the stress–strain curves produced provide additional information on the amount of strain energy involved in the process. A **metal wire** loaded beyond its elastic limit and then unloaded produces a stress–strain curve that shows **permanent deformation**. The area within the **loop** is the amount of strain energy needed to cause this permanent deformation and the wire is slightly longer, i.e. permanently stretched.

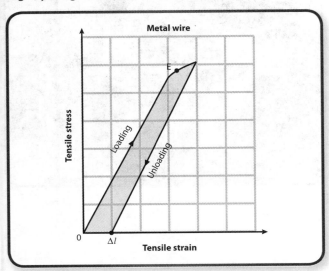

For polymer materials such as **polythene**, there is a very short elastic region but an exceptionally long plastic region. Provided that fracture has not occurred, unloading the polythene results in a significant permanent extension. The area within the loop, known as a **hysteresis**, is the work done or strain energy required to produce this permanent deformation. In such stress–strain graphs, it is usual to adopt the procedure of counting squares and to multiply this by the value for one square.

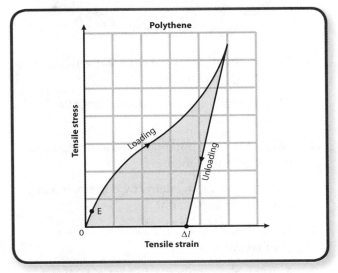

The same procedure can be adopted for other polymer materials such as **rubber**. However, in this case, provided that the rubber does not break, it will return to its original length and no permanent

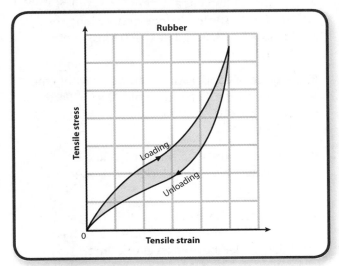

extension is produced. The reason for the difference between the loading and unloading curves is that some of the elastic strain energy is converted into **internal energy** of the molecules in the rubber when it is unstretched or unloaded. This usually appears as a temperature rise with the rubber becoming warmer when stretched and unstretched. The area of the hysteresis loop then gives the amount of internal energy retained by the rubber.

QUICK TEST

1. Give an equation for the work done in stretching a spring.

2. What is another term for elastic potential energy?

3. Why is the word potential used in the term for elastic potential energy?

4. How can the strain energy by determined in a force–extension graph?

5. What does the area under a stress–strain graph represent?

6. A spring has a stiffness constant of $50\,\mathrm{N\,m^{-1}}$ and is stretched elastically by 15 mm. Determine the strain energy stored in the spring.

7. A rubber band has a stiffness constant of $200\,\mathrm{N\,m^{-1}}$. The work done in stretching the band is 0.12 J. Calculate the extension of the rubber band.

8. A piece of copper wire is stretched within the elastic region. For a stress of 120×10^6 Pa, the strain is 1×10^{-3}. Determine the strain energy density.

9. What does the area within a loading–unloading curve for a length of rubber represent?

10. What does the area between loading and unloading a metal wire represent if it has been loaded beyond its elastic limit?

SUMMARY

- The energy stored in a stretched wire is called the **elastic potential energy or strain energy**

- The strain energy is given by $\frac{1}{2}F\Delta l$ or $\frac{1}{2}k(\Delta l)^2$ and has units of J or N m

- The strain energy is given by the area under a force–extension curve

- The strain energy density is the strain energy per unit volume and is given by $\frac{1}{2}$ stress $\times$ strain or the area under a stress–strain graph

- The loading and unloading **behaviour of materials can give rise to stress–strain curves containing hysteresis loops**

- The **area within these loops gives a measure of the energy required to permanently deform materials (e.g. metals, plastics) or the change in internal energy of a material (e.g. rubber)**

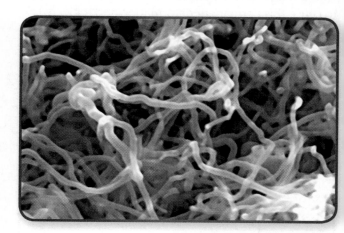

PRACTICE QUESTIONS

1. The graph shows the stress–strain curve for a metal wire. The cross marks the point of fracture.

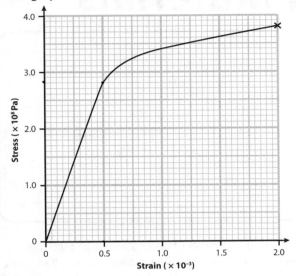

a) Calculate the Young modulus of the metal.
[3 marks]

b) State the ultimate tensile stress of the metal.
[1 mark]

c) Determine the energy stored per unit volume in the metal prior to fracture. **[3 marks]**

d) From the graph, state whether the metal is tough or brittle. Explain the reason for your choice.
[2 marks]

2. An experiment is carried out to determine the stiffness constant of a wire. The table below shows the results of this experiment.

Load (N)	0	1	2	3	4	5	6	7	8	9	10
Extension (mm)	0	0.7	1.3	1.8	2.2	2.9	3.5	4.1	4.7	5.6	6.3

a) Plot a graph of load against extension and determine the stiffness constant of the wire. **[6 marks]**

b) By considering the work done in stretching the wire, show that the energy stored is given by $\frac{1}{2}F\Delta l$, where F is the load and Δl is the extension. **[2 marks]**

c) Calculate the energy stored in the wire up to its maximum extension of 10 mm. **[2 marks]**

3. The graph shows the stress–strain behaviour of spider silk.

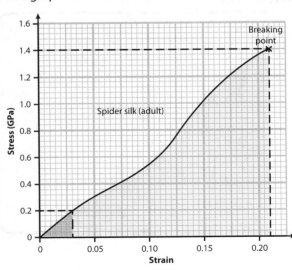

a) Calculate the Young modulus of spider silk when it is initially stressed. **[2 marks]**

b) Hence determine the strain energy density up to this point. **[2 marks]**

c) With reference to the graph, explain the behaviour of spider silk beyond its elastic limit. **[2 marks]**

d) Determine the strain energy density up to the maximum strain. **[3 marks]**

e) Hence determine the amount of elastic energy stored in a piece of spider silk 50 mm long and 4.0 μm in diameter, when strained to the maximum. **[4 marks]**

Microstructure of Materials

The Microstructure of Materials

The bulk properties of materials can be explained by looking at their **microscopic structure**. Solid materials tend to be grouped into four broad categories: **crystalline solids**, **amorphous solids**, **glasses** and **polymers**. The bulk properties of these materials, such as mode of fracture, Young modulus and plastic deformation, can be related to their microstructure and in particular to defects within their structure. **Metals** are crystalline structures with a regular, highly ordered array of ions. A pure metallic crystal, containing few defects, is theoretically the strongest a material can be. However, the process of manufacture, which involves **heating**, **rolling** and **cooling**, induces significant mechanical defects within the material. Such defects, which are known as **point**, **line** and **planar defects**, dominate the mechanical and physical properties of the metal.

Dislocations and Grain Boundaries

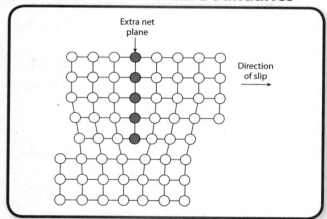

In metals, the mechanical properties are determined by the number and movement of **line defects** called **dislocations**. It is the presence of dislocations that make metals weak. The ductility of metals was originally thought to be due to planes of atoms sliding past each other. However, it is now known that it is the **movement of dislocations** that produces **slip** and gives rise to **plastic behaviour**. In most metals dislocations move very easily and there are many of them. In order to strengthen a metal, the movement of dislocations needs to be hindered and this is achieved through **mechanical working**, **heat treatment** and **alloying**.

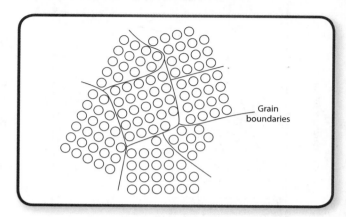

Increasing the number of dislocations and allowing them to obstruct each other's movement is called **work hardening**, which is achieved through rolling, extruding and hammering. These treatments, as well as quenching (heating and/or rapid cooling), also tend to break up the crystal structure into smaller crystals or **planar defects** called **grains**. These can range in size from nanometres to millimetres and each grain has a slightly different lattice orientation from its neighbour. It is the resulting **grain boundaries** that trap dislocations and prevent them from moving further through the metal. Heating the metal to high temperatures allows the untanglement of dislocations.

Point Defects

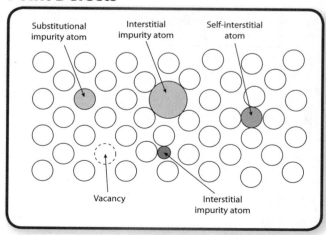

The introduction of **impurities** into a metal, such as through **alloying**, leads to the crystal lattice being distorted locally. These **point defects** include **interstitial impurity atoms** (e.g. adding carbon into an iron lattice to make steel), **substitutional impurity atoms** (e.g. zinc atoms replacing copper atoms to form brass) and **self-interstitial atoms** or **vacancies** that causes localised distortions within the lattice. All of these inhibit the movement of dislocations and hence strengthen the metal.

Ceramics

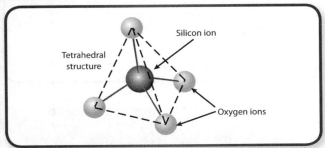

Tetrahedral structure

Silicon ion

Oxygen ions

Ceramics are also crystalline to a large extent but do not possess movable dislocations, so they are considerably stronger than metals. Ceramic materials are based largely on **ionic bonding** between oxygen and metals such as aluminium, magnesium and silicon in a **tetrahedral structure**, and **covalent bonding** between non-metal atoms. The crystal structure of quartz (shown) is a continuous framework of SiO_4 (silicon–oxygen) tetrahedra forming a very rigid structure. Such a structure is used to bond quartz crystals in the **long silica tetrahedral chains** found in asbestos or the **sheets of tetrahedra** found in mica. Clays such as kaolinite, consisting of silica (silicon oxide, SiO_2) and aluminium ceramics, form a layered silica structure with plates that easily slide over each other when wet but when heated form a very rigid material.

Amorphous Solids

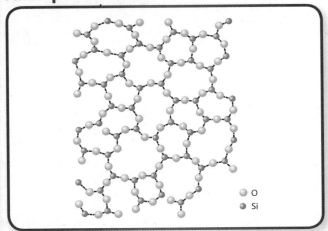

- O
- Si

Amorphous solids such as **glasses** do not possess a regular structure at all and are in fact a random array of molecules including silica mixed with other oxides that have been 'frozen'. In this sense, they are considered to be **supercooled liquids**; they have no long-range order in the way that the silica tetrahedra are arranged. Drawn glass is very strong but surface scratches and flaws makes the glass weak when under tension, which allows **cracks** to develop where concentrates. The increase in energy due to new surfaces being generated propagates the crack through the amorphous lattice which ruptures, often with dramatic consequences. Resistance to the spreading of cracks is called **toughness** and this can be measured from the strain energy stored in the glass using a force–extension or stress–strain graph.

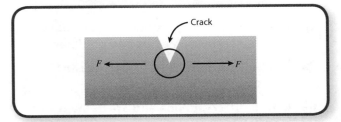

Crack

F ← → F

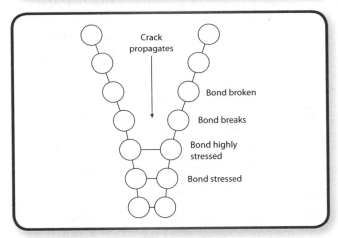

Crack propagates

Bond broken

Bond breaks

Bond highly stressed

Bond stressed

Polymers

Polymeric materials possess both **crystalline** as well as **glassy** or **amorphous** properties and are formed by long-chain molecules. Glassy polymers include PVC and perspex whereas crystalline polymers include polyethylene and nylon. The strength of polymers depends on how the long chains interlock, i.e. on how much **cross-linking** takes place between the chains. The properties of a polymer are therefore very sensitive to its structure, which in turn is sensitive to its temperature. Because of this sensitivity, very few polymers obey Hooke's Law and their Young modulus varies widely with temperature. The only exception is **rubber**, which has the unusual behaviour of being elastic for very large strains. The cross-sectional area of rubber changes significantly for large tensile strains and this affects the tensile stress. Because of this, rubber appears to be stiffer than it really is.

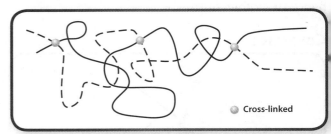

● Cross-linked

SUMMARY

- **Materials can be grouped into four classes: metals, ceramics, glasses and polymers**

- **The bulk properties of a material can be related to its microstructure**

- **Metals are crystalline and are defined by a rigid lattice framework; their mechanical properties are dominated by the movement and number of dislocations**

- **Materials that contain line defects such as dislocations allows plastic deformation (slip) to occur**

- **The movement of dislocations can be inhibited by introducing point defects (e.g. substitutional and interstitial impurities) and planar defects (e.g. grain boundaries)**

- **Ceramics do not possess dislocations**

- **Glasses are supercooled liquids; they are a combination of silica tetrahedral and other oxides and they have no long-range order**

- **Glasses are weak because of surface defects and flaws that promote crack propagation**

- **Polymers are made from long-chain molecules; they can have both crystalline and glassy properties**

QUICK TEST

1. Why are metals generally 'tough'?

2. What is meant by an amorphous structure?

3. What are polymers made from?

4. What makes a polymer stronger?

5. Why are ceramic materials stronger and stiffer than metals?

6. Why can metal atoms slide past each other relatively easily?

7. What is meant by work hardening?

8. Why are ceramics brittle?

9. Why are metals ductile?

10. Why are glasses called supercooled liquids?

11. Name three mechanisms that inhibit the movement of dislocations.

PRACTICE QUESTIONS

1. The diagram shows the stress–strain curves for two materials, A and B.

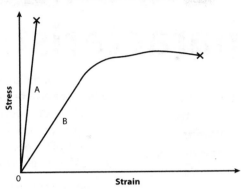

a) Give an example of what each of these materials could be. **[2 marks]**

b) With reference to your answer in part **a)**, explain the difference between:

    (i) elastic deformation and plastic deformation **[4 marks]**

    (ii) ductile and brittle materials. **[4 marks]**

2. The diagram shows the movement of a dislocation through a crystal.

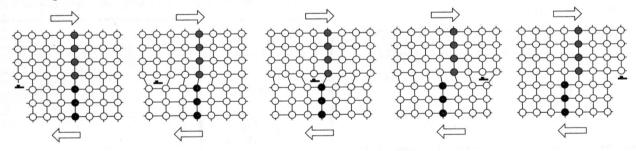

a) What is meant by a dislocation? **[2 marks]**

b) Explain how, under stress, the movement of a dislocation produces slip between crystal planes. **[3 marks]**

c) Explain two methods that could be used to inhibit the movement of dislocations.
Give examples of each method in your answer. **[4 marks]**

3. With the aid of a suitable graph and a comparison table showing examples, describe the main differences in microstructure between a metallic material and a ceramic. You should include in your description such features as: the Young modulus; elastic deformation; plastic deformation; and fracture. **[6 marks]**

Direct Current, Ohm's Law and I–V Characteristics

Current

The flow of charged particles, i.e. **electrons**, that move through a wire will give rise to a **current** (I) given by $I = \frac{\Delta Q}{\Delta t}$. If the charge, Q, has units of **coulombs** (C) and time, t, is in seconds then the current is defined in terms of **amperes** (A).

The charge can be calculated as current × time, i.e. $\Delta Q = I \Delta t$, which is often written as $Q = It$. The charge here is the total charge, i.e. the number of charge carriers (n) × the fundamental unit of charge (q), or $Q = nq$. For electrons, $q = 1.60 \times 10^{-19}$ C. In circuit diagrams the current flow is indicated by an arrow coming from the positive side of the cell or battery. This is the accepted convention and is called the **conventional current**.

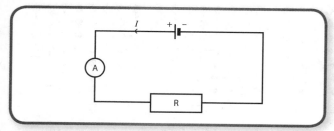

Current in a circuit is measured using an **ammeter** connected in **series** with a source (cell or battery) and the component, e.g. resistor, lamp, thermistor, diode, etc. Modern digital ammeters based on integrated circuits or older analogue ammeters may be used to measure current. Both devices possess resistance and so will affect the current in the circuit. To reduce this, ammeters are designed to have very low resistances and are calibrated to ensure readings reflect the true value of the current.

Potential Difference and Resistance

Electrons need to flow through the wire and this means that work has to be done on the electrons. This is usually supplied by the cell or battery, the **source**. The work done (W) per unit charge (Q) is called the **potential difference** (V), given by $V = \frac{W}{Q}$ or $W = QV$. The units of potential difference are **volts** (V). The potential difference is a measure of the amount of energy per unit charge converted to other forms of energy in a component, so $V = \frac{E}{Q}$. Since $Q = It$ and **power** is equal to the rate of work done, i.e. $P = \frac{W}{t}$, it follows that $V = \frac{P}{I}$ or $P = IV$, where power is measured in watts.

When any component is placed in an electrical circuit, it presents a barrier to the flow of charge through the circuit. Each component is said to have a **resistance**, R, and this can be expressed as a ratio of the potential difference to the current, $R = \frac{V}{I}$; resistance is measured in **ohms** (Ω). A component has a resistance of 1 Ω if a potential difference of 1 V makes a current of 1 A flow through it. Using this relationship provides an alternative way to express electrical power, giving two further useful equations, i.e. $P = I^2 R$ and $P = \frac{V^2}{R}$. When current passes through a resistor, it has to do work and this is dissipated as energy in the form of heat known as **Joule heating**.

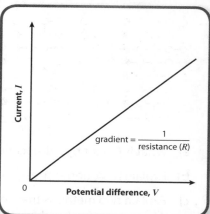

The relationship $R = \frac{V}{I}$ or $V = IR$ means that, for constant resistance R, $I \propto V$, and this is known as **Ohm's Law**. The resistance is only constant provided that the temperature remains constant and, under these conditions, the current is directly proportional to the potential difference across the resistor. We call materials that exhibit such behaviour **ohmic conductors**. Most **metals** are ohmic conductors at room temperature but, in practice, most components heat up as current passes through them.

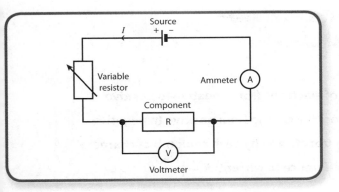

Current–Voltage Graphs

Current–voltage graphs are a useful way of recording the behaviour of a conductor. The circuit shown is one example of how the I–V behaviour can be observed. For an I–V graph, the gradient at a point represents $\frac{1}{R}$. For a piece of metallic wire at constant temperature, the I–V response is linear. If the temperature is increased then the resistance of a metal wire increases (and $\frac{1}{R}$ decreases) as the charge carriers, the **conduction electrons**, make more collisions with the vibrating atom (ions). The metal is said to have a **positive temperature coefficient (ptc).** If the temperature rises significantly the I–V behaviour of a wire shows the curvature expected. A filament lamp is an ideal example of this.

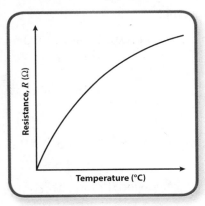

Thermistors and **light-dependent resistors (LDRs)** are devices that are made from **semiconductor materials**. Their I–V profile indicates ohmic behaviour at constant temperature or constant lighting conditions but shows a steepening of the line with increasing temperature or brightness, respectively indicating that the resistance is lowered; this is opposite to metallic behaviour and arises because the number of charge carriers increases when the temperature or light intensity increases. Such devices are said to have a **negative temperature coefficient (ntc).** Thermistors are useful for controlling central heating systems and fridges, while LDRs are useful as exposure meters for cameras and as light sensors.

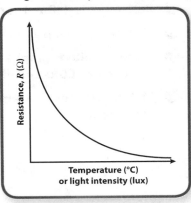

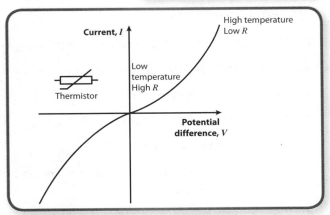

Diodes including **light-emitting diodes (LEDs)** are semiconductor devices that permit the flow of current in one direction only, a condition known as **forward bias**. When the voltage is reversed (**reverse bias**), there is negligible current, as shown in the I–V graph. After a certain voltage, their behaviour follows a normal ohmic relationship.

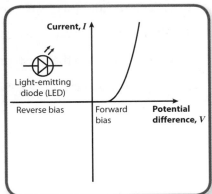

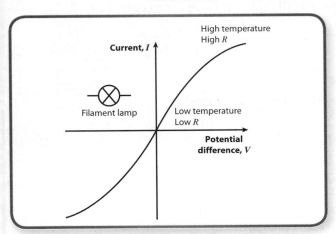

SUMMARY

- Electric current **is the flow of charge:** $I = \frac{\Delta Q}{\Delta t}$
- **Current** in a circuit is carried by the movement of electrons from negative to positive
- **Conventional current** is said to be the flow of positive charge from positive to negative
- **Potential difference** measures electrical energy transferred by each coulomb of charge: $V = \frac{E}{Q}$
- **Resistance** is defined as the ratio of potential difference to current: $R = \frac{V}{I}$
- **Electrical power** is $P = IV = I^2 R = \frac{V^2}{R}$
- **Current–voltage** graphs show the behaviour and characteristics of devices such as filament lamps, thermistors, LDRs and diodes
- **Metals** have positive temperature coefficients **and** semiconductors **have** negative temperature coefficients

QUICK TEST

1. What is a cell?

2. What is the relationship between current, charge and time?

3. How much charge flows through a torch bulb when a current of 0.025 A flows for 3 minutes?

4. What is meant by conventional current?

5. What is potential difference?

6. How is resistance defined?

7. The current in an LED is 30 mA when there is a potential difference across it of 9 V. Determine the resistance of the diode.

8. What is the power of a 12 V bulb that has a current of 2.5 A flowing through it?

9. What is the electrical unit for (i) electric charge and (ii) electric current? Choose one each from $J\,C^{-1}$, $J\,s^{-1}$, $C\,s^{-1}$, $A\,V^{-1}$ and $A\,s$.

10. What is meant by a positive temperature coefficient?

11. What is meant by an ohmic conductor?

12. What happens to the resistance of an LDR when the light intensity increases?

13. What is the main characteristic of a diode?

PRACTICE QUESTIONS

1. a) Sketch a graph of the *I–V* characteristics of an LDR (only for positive values of *I* and *R*), indicating its response in the dark and light. **[2 marks]**

b) Sketch a graph to show how the resistance in the LDR varies as a function of the light intensity. **[2 marks]**

c) Explain the reasons why when the light intensity diminishes the resistance increases. **[3 marks]**

2. The table shows the results of measuring the current through an electronic component.

Current, *I* (A)	0	0.003	0.007	0.010	0.013	0.017	0.020	0.023	0.026
Voltage, *V* (V)	0	1.0	2.0	3.0	4.0	5.0	6.0	7.0	8.0

a) Plot a graph to show the current–voltage characteristics of this component. **[3 marks]**

b) What does the value of the slope of the graph represent? **[1 mark]**

c) Determine the resistance of the component. **[2 marks]**

d) Give a suggestion as to what the component could be and explain its behaviour. **[2 marks]**

e) The temperature of the component is now raised significantly and kept constant at this elevated temperature; the experiment is now repeated. What result would you expect to obtain under these conditions? **[2 marks]**

f) Explain the reasons why it might show this characteristic. **[3 marks]**

3. a) Explain whether the *I–V* characteristics of a filament lamp represent an ohmic conductor. **[4 marks]**

b) Sketch the *I–V* characteristics of a filament lamp. **[2 marks]**

c) Does the resistance of the metal conductor increase or decrease with temperature? Explain your answer. **[3 marks]**

d) Describe how you would determine the *I–V* characteristics of a filament lamp; include a circuit diagram in your answer. **[3 marks]**

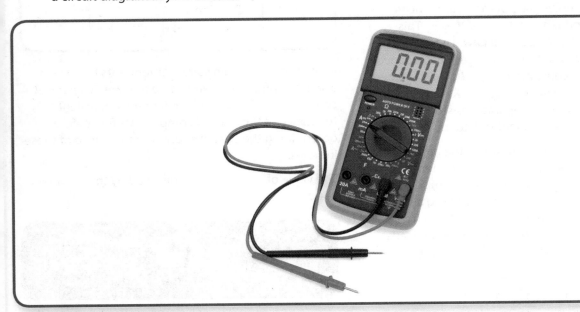

Resistance, Resistivity, Conductivity and Superconductivity

Resistances in Series and in Parallel

When conducting electricity, all materials and components have some **resistance** (R) to the flow of charge. Components are linked within a circuit in one of two ways: **in series** or **in parallel**. For resistances **in series**, the current is identical at all points throughout the circuit (there are no junctions) but the potential difference is divided between the components so that the source or input potential difference

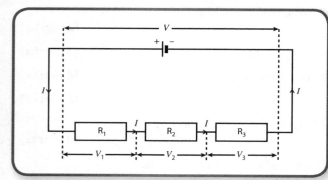

is the sum of the individual potential differences across each component, i.e. $V = V_1 + V_2 + V_3$. This is known as **Kirchhoff's Second Law**. The voltage splits proportionally to the resistance in accordance with $V = IR$. As I is constant, $IR_{total} = IR_1 + IR_2 + IR_3$, which leads to the result that $\mathbf{R_{total} = R_1 + R_2 + R_3}$. This means that the three separate resistors can be replaced by a single equivalent resistor of value R_{total}.

For circuits with resistors **in parallel**, the potential difference is identical across each component but the current is split at each junction (each loop) so that $I = I_1 + I_2 + I_3$, i.e. the total current entering a junction is equal to the total current leaving that junction. This is **Kirchhoff's First Law**. The above equation leads to the result that $\frac{V}{R_{total}} = \frac{V}{R_1} + \frac{V}{R_2} + \frac{V}{R_3}$ and therefore the total resistance is given by

$$\frac{1}{R_{total}} = \frac{1}{R_1} + \frac{1}{R_2} + \frac{1}{R_3}$$

i.e. the reciprocal of the combined resistance is the sum of the reciprocals of all the individual resistances. Again, all three resistances can be replaced by a single equivalent resistor of value R_{total}. For two identical resistors in parallel, the combined resistance is equal to half the value of each one, and any parallel

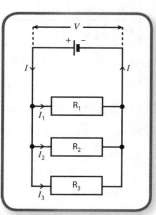

combined resistance is always less than the value of the smallest individual resistance.

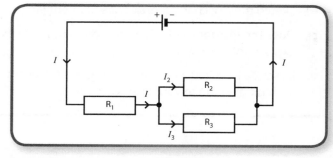

Many circuits contain some components that are in a series configuration and some that are in a parallel combination. It is sensible to find the equivalent resistance of the parallel components first before determining the overall equivalent resistance of those in series.

The value of a resistor is determined by the coloured bands.

Determining Resistance

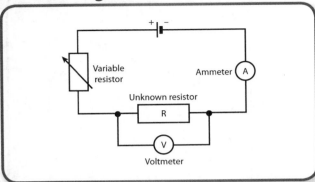

The value of the resistance of an unknown resistor can be found using the circuit shown. The voltmeter measures the potential difference across the resistor. The variable resistor and the source potential difference can be used to adjust the current in precise steps and the current and voltage across the resistor recorded to produce a current–voltage or voltage–current graph. The gradient of the line of an I–V or V–I graph gives the value of $\frac{1}{R}$ or R, respectively.

Resistivity and Conductivity

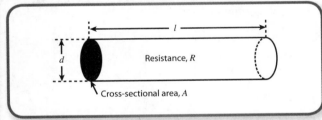

The resistance of a component depends on a number of physical factors. For the simple case of a piece of wire, the resistance depends on:

- the **length** (l) of the wire; the longer the wire, the more difficult it is for electrons to flow through it
- the **cross-sectional area** (A) of the wire; the broader the wire, the easier it is for electrons to flow through the wire.

Combining these effects gives $R \propto \frac{l}{A}$ and hence $R = \rho\frac{l}{A}$, where the constant of proportionality, ρ, depends on the atomic structure of the wire material as well as on environmental effects such as temperature. This constant ρ is called the **resistivity** of the material. The resistance of a wire depends on its dimensions but its resistivity depends on what the material is made from. Resistivity is therefore given by $\rho = \frac{RA}{l}$, with units of $\Omega\,$m. The lower the resistivity of a material, the better it is at conducting electricity. Resistivity can be measured using a **potential divider** circuit similar to that below where a fixed current value allows the potential difference to be measured across different lengths of wire. Resistance can be plotted against length and the gradient gives a value of $\frac{\rho}{A}$, from which the resistivity can be determined. Sometimes the reciprocal of resistivity is used, and this is called **conductivity** (σ): $\sigma = \frac{Gl}{A}$, where G is the reciprocal of resistance ($G = \frac{1}{R}$) and called the **conductance**. The unit of conductance is the **siemens** (S), so the units of conductivity are $S\,m^{-1}$. However, resistivity is the most common parameter used when describing properties of materials.

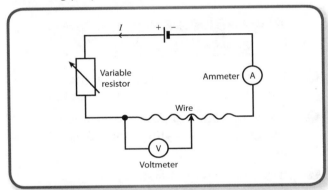

Material	Resistivity ($\Omega\,$m)
Metals (conductors)	10^{-7} to 10^{-8}
Semiconductors	10^{-2} to 10^{2}
Insulators	10^{14}

The resistivity of a material depends on its intrinsic properties and on the mobility of electrons to flow through the material, and is also dependent on temperature. With increasing themperature, the resistivity of metallic materials slowly increases whereas for semiconductor materials (such as silicon) it rapidly decreases, a property that is exploited widely in the electronics industry.

Superconductivity

Collisions between electrons and atoms in a material result in the transfer of electrical energy usually in the form of heat. The resistivity of many materials, particularly metals, can be reduced significantly by cooling them down to extremely low temperatures. Below a **critical** or **transition temperature**, the resistivity disappears completely and the material becomes a **superconductor**. Transition temperatures for common metals are below 10 K (−263°C); the highest critical temperature reached so far is ~150 K (−123°C). Any material with a critical temperature above 77 K (−196°C) is called a **high-temperature superconductor**. Applications of superconductors include power cable transmission and electromagnets.

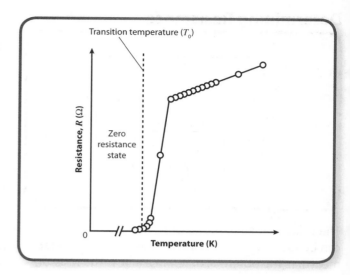

SUMMARY

- The conservation of current (electric charge) is known as **Kirchhoff's First Law**

- The conservation of potential difference (energy) is known as **Kirchhoff's Second Law**

- The equivalent resistance R_{total} of resistors connected in series is given by $R_{total} = R_1 + R_2 + R_3 + \dots$

- The equivalent resistance R_{total} of resistors connected in parallel is given by $\frac{1}{R_{total}} = \frac{1}{R_1} + \frac{1}{R_2} + \frac{1}{R_3} + \dots$ and then taking the reciprocal to obtain R_{total}

- **Resistance** of a wire depends on its dimensions

- **Resistivity** is a measure of a material's ability to oppose the flow of an electric current and is given by $\rho = \frac{RA}{l}$ and depends on the structure of the material

- The absence of a measurable electrical resistance below a critical temperature is called **superconductivity**

QUICK TEST

1. You are given three resistances in series of values 20 Ω, 100 Ω and 56 Ω. What is the value of the equivalent resistor that can replace these?

2. If the above three resistors were placed in parallel, what would the new equivalent resistor value be?

3. What is Kirchhoff's First Law?

4. What is the purpose of the variable resistor in a circuit that is used to measure the resistance of a component?

5. What are the three properties of a wire that its resistance depends upon?

6. How are resistance and resistivity related?

7. What are the units of resistivity?

8. What do we call a material that has no measurable resistance?

9. What do we call the temperature at which this occurs?

PRACTICE QUESTIONS

1. Two resistors of values 1 kΩ and 550 Ω are connected in parallel to a 12 V battery as shown in the figure.

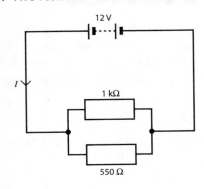

a) Calculate the overall resistance of this circuit (assume that the battery has negligible internal resistance). **[2 marks]**

b) Calculate the current drawn from the 12 V battery. **[2 marks]**

An additional 500 Ω resistor is now added to the circuit in series with the battery.

c) Calculate the overall resistance in this new circuit and determine the current drawn from the battery. Comment on what effect this additional resistor has on the current drawn from the battery. **[3 marks]**

2. A 200 Ω resistor and 500 Ω resistor are connected in series with each other. The series combination is connected in parallel with a 1 kΩ resistor and a 9 V battery of negligible internal resistance, as shown in the figure.

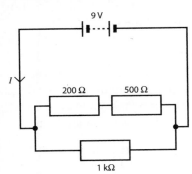

a) Calculate the total resistance of the circuit. **[2 marks]**

b) Calculate the current flowing in the circuit. **[2 marks]**

c) Determine the power supplied by the battery. **[1 mark]**

d) Determine the power supplied to the 1 kΩ resistor. **[1 mark]**

3. The following measurements were recorded in an experiment to determine the resistivity of a particular material. The wire had a diameter of 0.23 mm and a length of 1.02 m.

Potential difference across the material (V)	0.0	1.0	2.0	3.0	4.0	5.0
Current through the material (A)	0.00	0.07	0.16	0.24	0.32	0.43

a) Plot a graph of the current (*y*-axis) against the potential difference (*x*-axis). **[3 marks]**

b) Using the formula for resistivity and Ohm's Law, show that the current I varies with the potential difference as $I = \frac{A}{l\rho}V$, where l is the length of the material of cross-sectional area A, and ρ is the resistivity of the material. What does $\frac{A}{l\rho}$ represent? **[3 marks]**

c) Using the formula in part **b)** and the graph, determine the resistivity of the material. **[4 marks]**

d) Comment on the type of material it is most likely to be made from. **[2 marks]**

Emf and Internal Resistance

Emf and Internal Resistance

The amount of electrical energy transferred to each coulomb of charge in a cell or battery is called the **electromotive force** or **emf** (ε), measured in volts. $\varepsilon = \frac{E}{Q}$, where E is the amount of work done on the charge or the amount of energy per unit charge converted from chemical energy into electrical energy. Some of this energy is converted to heat inside the cell itself, in the same way that energy is lost inside any electrical component as a result of the collisions between the charge-carrying electrons and atoms. This resistance of the cell itself is known as **internal resistance**, the effect of which is that the energy supplied to the circuit (per unit charge) is less than the total emf produced by the cell. If a high-resistance voltmeter is connected across a cell or battery then negligible current flows, resulting in negligible heating of the cell, and an accurate value for the emf can be measured.

Terminal Potential Difference and Emf

To represent the internal resistance generated by a cell or battery, a tiny resistor is drawn next to the cell. These two symbols are then enclosed by either a circle or dashed box (as shown). In a circuit that has a **load resistance** (R), the potential difference across this is called the **terminal potential difference (tpd)** (V). If there were no internal resistance in the cell then the terminal potential difference would equal the emf. However, this is generally not the case and so **tpd ≤ emf**. The energy used in overcoming the internal resistance is called the **'lost volts'**. The internal resistance is the resistance of the cell itself and cannot be separated from it. Some questions may say "ignore the internal resistance of the cell"; in these circumstances, the terminal potential difference is taken as the same as the emf.

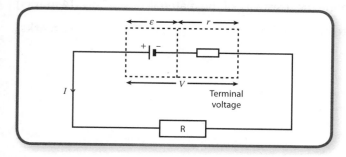

If R represents the equivalent (load) resistance of all the components in a circuit and r represents the internal resistance of the cell then the circuit can be drawn as shown. The total resistance of the complete circuit is then

$$R_{total} = R + r$$

and the current flowing is calculated using $V = IR$, i.e.

$$I = \frac{\varepsilon}{R_{total}} = \frac{\varepsilon}{R + r}$$

where ε is the cell's emf. Rearranging gives

$$\varepsilon = I(R + r) = IR + Ir$$

This equation states that the emf is the terminal potential difference (IR) plus the 'lost volts' (Ir) due to the internal resistance of the cell.

Emf and Power

Some energy (or power) is lost as heat when overcoming the internal resistance in the cell or other power supply. This can be calculated by using the equation $P = I^2R$, which when applied to the above equation for the emf gives

$$I\varepsilon = I^2R + I^2r$$

Here, $I\varepsilon$ is the **power supplied** by the cell, which is equal to the **power delivered to R** (I^2R) plus the **power wasted in the cell** owing to its internal resistance (I^2r). Since $I = \frac{\varepsilon}{R+r}$, the power delivered to the resistor R is given by $I^2R = \frac{\varepsilon^2R}{(R+r)^2}$ and this is a maximum when $R = r$, i.e. when a source delivers power to a load resistance (R), maximum power is only delivered when the load resistance is equal to the internal resistance of the cell or supply. In this case the load is said to be **matched** to the source.

Cells and batteries that are used in everyday devices such as torches, cameras and mobile phones have very small internal resistances, typically less than $1\,\Omega$. Car batteries also need to have low internal resistance in order to deliver a very high current. Power supplies, on the other hand, have very high internal resistance in order to make them safe if they suffered a short circuit. Under these circumstances, only a small current would flow since $I = \frac{V}{R}$. High-voltage supplies have been used extensively in oscilloscopes, 'old-style'

TV tubes and cathode ray devices, but are now being phased out for more advanced devices based on solid state technology.

Determining Emf and Internal Resistance

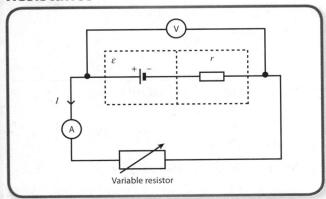

Variable resistor

The measurement of the emf and the internal resistance of a cell or battery can be made using a similar circuit to that used to measure the resistance of an unknown resistor. However, in the circuit shown, a **high-resistance voltmeter** is connected directly across the terminals of the cell to measure the terminal potential difference. The current is altered by using a variable resistor in series with the ammeter and cell.

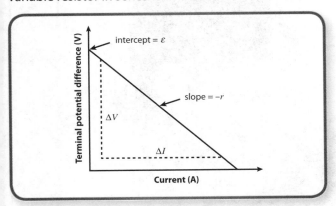

Plotting a graph of terminal potential difference V (volts) against current I (amperes) provides a measure of both the emf and the internal resistance. Manipulating the equation for the emf gives an alternative expression for the terminal potential difference as $V = -rI + \varepsilon$,

which can be directly compared with the equation for a straight line, $y = mx + c$. Note that the slope has a negative gradient equal to $-r$ where r is the **internal resistance**, and the **y-intercept** provides a direct measure of the **emf**.

Circuits in which there are more than one cell or battery connected in **series** have a net emf equal to the sum of the individual emf's provided that the cells are connected in the same direction. The total internal resistance is the sum of the individual internal resistances. Connection of cells in **parallel** results in an emf that is the same as for one cell but an internal resistance that is given by the equivalent resistance for resistances in parallel.

SUMMARY

- The **electromotive force (emf)** of a source measures the electrical energy gained per unit charge passing through the source

- The **potential difference** across the terminals of a source (the **terminal potential difference**) is **always less than the emf** of the source when the source is delivering a current

- For a source with emf ε and internal resistance r, $\varepsilon = \frac{I}{R+r}$, where R is the external circuit resistance and I is the current

- The **power supplied** by the cell is equal to the power delivered to R plus the power wasted owing to the cell's **internal resistance**

- **Maximum power** is delivered to a load resistance (R) when $R = r$

- The **terminal potential difference (tpd)** is given by $V = -rI + \varepsilon$. A graph of V against I gives a measure of both the emf and the internal resistance.

QUICK TEST

1. What is meant by electromotive force (emf)?

2. What unit is electromotive force measured in?

3. Why is the potential difference of a cell (or battery) less than its emf?

4. What is meant by terminal potential difference?

5. How can the internal resistance of a cell be measured?

6. When plotting a voltage–current graph for a cell, what do the gradient and y-intercept show?

7. What is the power supplied by the cell equal to?

8. Under what conditions does a cell or battery deliver maximum power to a circuit?

9. What is the total emf of two identical cells connected in the same direction when (i) in series (ii) in parallel?

10. A high-resistance voltmeter reads 15 V when connected across the terminals of a battery. This falls to only 13.5 V when the battery delivers a current of 0.3 A to a lamp. (i) What is the potential difference across the internal resistance? (ii) What is this called? (iii) Calculate the internal resistance of the battery.

PRACTICE QUESTIONS

1. A cell with an emf of 1.5 V has an internal resistance of 0.8 Ω. It is connected to a 5 Ω resistor that is in series with the battery.

    a) Draw a circuit diagram to represent the components in the circuit and show the direction of the conventional current. **[3 marks]**

    b) Calculate:

        (i) the current flowing in the circuit **[2 marks]**

        (ii) the terminal potential difference **[1 mark]**

        (iii) the power delivered to the 5 Ω resistor **[1 mark]**

        (iv) the power wasted in the cell. **[1 mark]**

2. A cell of emf 6.0 V and internal resistance 3.8 Ω is connected in series with another cell of emf 1.5 V and internal resistance 1.0 Ω. This combination is then linked in series with a 55 Ω resistor.

    a) Draw a circuit diagram to show this arrangement. **[2 marks]**

    b) Calculate the total resistance of this circuit. **[1 mark]**

    c) Determine the current flowing through the 55 Ω resistor. **[2 marks]**

    d) Calculate the potential difference across the 55 Ω resistor. **[2 marks]**

3. A series of measurements was made to determine the emf and internal resistance of a cell using an ammeter in series with a variable resistor. The results are shown in the voltage–current graph.

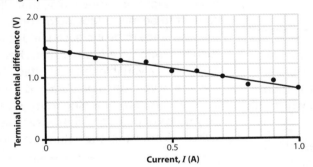

    a) Rearrange the expression $\varepsilon = IR + Ir$ to make the terminal potential difference the subject of an equation that will allow the above graph to be interpreted. **[4 marks]**

    b) Using this equation and the graphical data shown, determine:

        (i) the emf of the cell

        (ii) the internal resistance of the cell. **[3 marks]**

The Potential Divider and its Applications with Sensors

The Potential Divider

A **potential divider** in its simplest form is a circuit consisting of two resistors, R_1 and R_2, in series with a source of fixed potential difference, the **source voltage** or **input voltage**, V_{in}. The potential difference of the source is then shared (between V_1 and V_2) across the two resistors in the ratio of their resistance values, i.e.

$$\frac{V_1}{V_2} = \frac{R_1}{R_2}$$

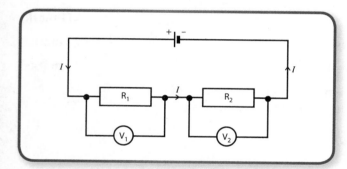

Potential dividers can therefore be used to provide a variable potential difference between zero and the source potential difference; as a consequence, there are a large number of useful applications to which this can be applied.

In its simplest form, the potential divider has an input voltage V_{in} and two fixed-value resistors R_1 and R_2 connected in series; the total resistance in the circuit (ignoring the internal resistance of the source) is then $R = R_1 + R_2$. Using $V = IR$ (remember that the current flowing is the same everywhere in a series circuit), the total voltage across the resistors is $V_{in} = I(R_1 + R_2)$. Rearranging this gives $I = \frac{V_{in}}{R_1+R_2}$. The potential difference across R_1 is just IR_1 and substituting for I gives a value for the **output voltage** or **output potential**, V_1, as

$$V_1 = V_{in}\left(\frac{R_1}{R_1 + R_2}\right)$$

The corresponding output voltage at R_2 is then given by the expression

$$V_2 = V_{in}\left(\frac{R_2}{R_1 + R_2}\right)$$

Remember that the output voltage across a resistor is the input voltage multiplied by the ratio of that particular resistance divided by the total resistance in the circuit. If the resistor values are identical then the output voltage is simply halved. The equations can easily be extended to more than two resistors provided that the sum of all the resistors is used as the denominator.

The Potentiometer

The two resistors above may be replaced by a variable resistor called a **potentiometer**. These come in various shapes and sizes but they are simply a piece of wire, either straight or semi-circular, that allows a sliding contact to be made. Moving or rotating a slider allows the resistance to be increased or decreased. Using such potentiometers (also called **rheostats**) in this way allows a variable output voltage between 0 and the source voltage to be obtained. There are numerous applications of this ranging from controlling the brightness of bulbs and lights (such as in a dimmer switch) to adjusting the volume in audio equipment.

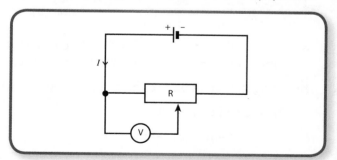

A simple potentiometer can be made from a length of high-resistance metal wire, such as constantan or nichrome, stretched along and attached to a metre rule. A metal probe is then used to make contact with the wire at various positions along its length and the output voltage measured by a voltmeter (as shown in the circuit). A graph can be plotted of the potential difference (along the y-axis) against the length of wire (along the x-axis) and the best straight line drawn.

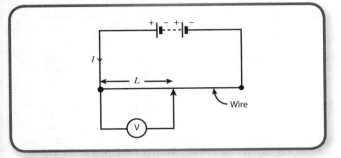

potential difference is required following an increase in temperature so it is usual to connect the voltmeter across the fixed resistor as shown. Applications are again numerous, ranging from central heating thermostats and freezer controls to fire alarm devices.

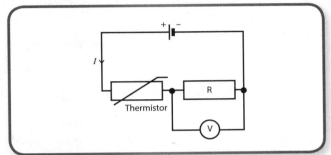

Applications of the Potential Divider

Another application of potential dividers is using them as **electronic light and temperature sensors**. Both light-dependent resistors (LDRs) and thermistors are made from semiconductor materials and hence have negative temperature coefficients. For an **LDR**, this means that it can change its high resistance value, of the order of mega-ohms in the dark, to only a few hundred ohms in the light. This degree of sensitivity makes it a very useful device as a light sensor. If an LDR is connected in series with a fixed or variable resistor, replacing R_1 or R_2 in the original potential divider circuit on page 116, then as the light intensity increases the output voltage across the LDR will decrease, which provides a variable voltage that can be used as a simple on/off switch.

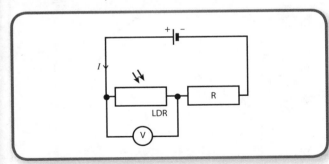

A very similar response can also be obtained using a **thermistor**. This is a device that possesses a very high resistance value at low temperatures and a low resistance value at high temperatures. Again, by replacing one of the fixed resistors with a thermistor, the output voltage across the fixed resistor will vary depending on whether the temperature increases or decreases. In most applications, the increase in

SUMMARY

- A **potential divider** **shares the source potential difference between two resistor components in the circuit in the ratio**
  $$\frac{V_1}{V_2} = \frac{R_1}{R_2}$$

- The **output voltage across a particular fixed-value resistor (say R_2) is**
  $$V_{\text{out}} = V_{\text{in}} \left(\frac{R_2}{R_1 + R_2} \right)$$

- The **fixed resistors can be replaced by a potentiometer that allows for a continuously variable output voltage ranging from 0 up to the value of the input voltage**

- A **simple potentiometer can be made from a length of nichrome or constantan wire using a simple circuit involving an input source and a voltmeter**

- One of the fixed resistors in a potential divider circuit can be replaced by a **thermistor that provides a variable output voltage that depends on temperature; this results in a temperature-sensitive sensor with many applications**

- One of the fixed resistors can be replaced by a **light-dependent resistor that provides a changing output voltage due to changing light conditions; such devices can be used in a range of light sensor applications**

QUICK TEST

1. What does a potential divider provide?

2. In a potential divider circuit, are the resistor components placed in series or in parallel with the source potential difference?

3. If two resistors of equal value are placed in the circuit, what is the value of the potential difference across each of the resistors?

4. If the value of the resistances are $R_1 = 5\,\Omega$ and $R_2 = 10\,\Omega$, what is the ratio of their potential differences, i.e. $\frac{V_1}{V_2}$?

5. In the above example, if the source potential difference is 3 V, what is the potential difference across R_2?

6. In what type of circuit do potential dividers have numerous applications?

7. What two types of resistors are used to control temperature and light intensity?

8. What is normally included in such devices that allows a switch to be made?

9. Give two household examples of using a potential divider circuit.

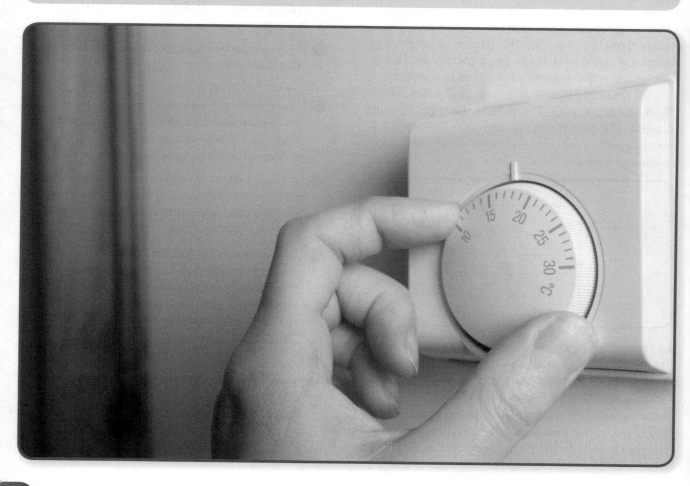

PRACTICE QUESTIONS

1. In the circuit shown, the 12V battery has negligible internal resistance.

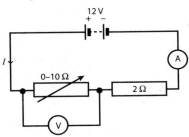

a) If the variable resistor is set to 2 Ω, what is the voltage across it? [1 mark]

b) It is now reset to 8 Ω. What is the new value of the voltage across it? [2 marks]

c) Calculate the current in the circuit when the variable resistance is 8 Ω. [2 marks]

d) If the voltmeter is now replaced by a 6 Ω resistor and the variable resistor is set at 6 Ω, what is the value of the current in the circuit? [3 marks]

2. A temperature sensor system consists of a 6.0V battery connected in series with a low-resistance ammeter, a resistor R_1 and a thermistor. The resistance of the thermistor at 24°C is 1.2 kΩ and the voltmeter across R_1 gives a reading of 2.2V.

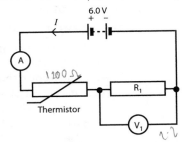

a) Determine the potential difference across the thermistor. [1 mark]

b) Calculate the resistance R_1. [2 marks]

c) Calculate the current through the thermistor. [2 marks]

d) If the temperature starts to rise rapidly, what effect would this have on the thermistor (which has a negative temperature coefficient)? [1 mark]

e) What would happen to the voltage across the resistor? Explain your reason. [2 marks]

3. A light-sensing circuit uses an LDR connected in series with a 75 kΩ resistor, a 9V power supply and a low-resistance micro-ammeter.

a) Draw a circuit diagram showing all of the above components. [3 marks]

b) When it is dark, the LDR has a resistance value of 180 kΩ. Determine the current in the circuit under these conditions. [3 marks]

c) Calculate the potential difference across the LDR under the same conditions given in (b). [2 marks]

d) In full daylight, the resistance of the LDR falls to 10 kΩ. Determine the current now flowing through the micro-ammeter and the potential difference across the LDR. [3 marks]

e) Determine the resistance of the LDR when the potential difference across it is 3 V. [2 marks]

Answers

Day 1

Base Units, Prefixes and Estimation
QUICK TEST (page 6)

1. kilogram, second, ampere, metre
2. $p = mv = $ mass $\times$ velocity $= kg \times m\,s^{-1} = kg \times m\,s^{-2} \times s = Ns$
3. 8×10^{-19} J
4. **(i)** 7.79×10^9 km **(ii)** 51.9 AU
5. **(i)** 1.4 fm **(ii)** 1.4×10^{-6} nm
6. 6.35×10^{-7} m
7. 2.4×10^{-5} m
8. $E_p = 10^3 \times 10 \times 10 = 10^5$ J
9. mass $\times$ acceleration due to gravity $\times$ height

$$= kg \times m\,s^{-2} \times m$$
$$= kg \times m^2\,s^{-2}$$
$$= kg\left(m\,s^{-1}\right)^2 = J$$

PRACTICE QUESTIONS (page 7)

1. **a)** $E_k = qV$
   (i) $E_k = 1.60 \times 10^{-19} \times 5 \times 10^3$
   $= 8.0 \times 10^{-16}$ J **[1 mark]**
   (ii) 5×10^3 eV or 5000 eV **[1 mark]**
   b) $n = 3 \times 10^{12}$ electrons
   $E_{total} = nE_k$ **[1 mark]**
   $= 3 \times 10^{12} \times 8 \times 10^{-16}$ J
   $= 2.4 \times 10^{-3}$ J **[1 mark]**
   $= 2.4$ mJ **[1 mark]**
   c) 20% absorbed $\rightarrow$ 80% reach the detector
   $E_{total} = 2.4 \times 10^{-3} \times 0.8$ **[1 mark]**
   $= 1.92 \times 10^{-3}$ J or 1.92 mJ **[1 mark]**

2. **a)** $d = 2.3$ mm
   $A = \pi r^2 = \frac{\pi}{4} d^2$ **[1 mark]** $= \frac{\pi\left(2.3 \times 10^{-3}\right)^2}{4}$
   $A = 4.2 \times 10^{-6}$ m^2 **[1 mark]**
   b) $E = 210$ GPa
   (i) 2.1×10^{11} Pa **[1 mark]**
   (ii) $210\,000 \times 10^6$ Pa $= 210\,000$ MPa **[1 mark]**
   c) $l = 4.5$ m, $F = 50$ N
   $$E = \frac{Fl}{Ae} \rightarrow e = \frac{Fl}{AE} \text{ [1 mark]}$$
   (i) $e = \dfrac{50 \times 4.5}{\left(4.15 \times 10^{-6}\right)\left(2.1 \times 10^{11}\right)} = 2.58 \times 10^{-4}$ m
   $\approx 2.6 \times 10^{-4}$ m **[1 mark]**
   (ii) $e = 0.26$ mm **[1 mark]**
   d) (i) If F is doubled, e is doubled **[1 mark]**
   (ii) If A is doubled, e is halved **[1 mark]**
   as E is a constant **[1 mark]**

3. **a)** $R = \dfrac{\rho l}{A} \rightarrow \rho = \dfrac{RA}{l}$ **[1 mark]**
   $\rho \Rightarrow \Omega \cdot \dfrac{m^2}{m} \Rightarrow \Omega\,m$ **[1 mark]**

b) Since $A = \pi\left(\dfrac{d}{2}\right)^2$, $A \propto d^2$ **[1 mark]**
   If d is doubled, A is increased by a factor of 4 **[1 mark]**
   Since $R \propto \dfrac{1}{A}$, R is decreased by a factor of 4 **[1 mark]**
c) $R = \dfrac{\rho l}{A} = \dfrac{1.7 \times 10^{-8} \times 0.8}{3.2 \times 10^{-8}}$ **[1 mark]**
   $\approx \dfrac{10^{-8} \times 1.0}{10^{-8}} \approx 1\,\Omega$ **[1 mark]**
d) $R = \dfrac{1.7 \times 10^{-8} \times 0.8}{3.2 \times 10^{-8}} = 0.425 \approx 0.43\,\Omega$ **[1 mark]**
e) If ρ increases then, since $R \propto \rho$, R also increases **[1 mark]**

Physical Measurements
QUICK TEST (page 10)

1. resolution 0.01 s; precision is ± 0.005 s (accept ± 0.01 s to ± 0.02 s)
2. ruler is truly vertical; avoid error due to parallax
3. micrometer; ± 0.01 mm
4. 130 ± 20
5. 11.4 ± 0.3
6. (3.36 ± 0.01) V
7. (4.3 ± 0.1) A
8. (0.654 ± 0.001) m

PRACTICE QUESTIONS (page 11)

1. **a)** micrometer **[1 mark]**; resolution = 0.01mm **[1 mark]**; precision = ± 0.005 mm **[1 mark]**
   b) **Any one from:** it should be calibrated; zero error determined; use of ratchet to avoid deforming the wire; measuring at several different places along the wire **[1 mark]**
   c) True value

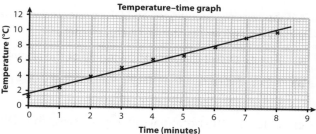

   Results are precise **[1 mark]** but not accurate **[1 mark]**
2. **a)** resolution = 0.1°C **[1 mark]**
   uncertainty = ± 0.05°C [accept ± 0.1°C] **[1 mark]**
   b) correct plot **[1 mark]**; axes labelled **[2 marks]**
   c) best line of fit drawn **[1 mark]**

Temperature–time graph

   d) Random errors are small **[1 mark]**; points plotted around best line of fit **[1 mark]**; systematic error identified; zero error on thermometer **[1 mark]**; line should pass through origin **[1 mark]**

3. a) Mean value is 792 mm [accept 791.6 mm] **[1 mark]**

   b) Not accurate **[1 mark]**; true reading is 795 mm and all five measurements are either 791 or 792 mm, i.e. 4 or 3 mm away from the true value **[1 mark]**

   c) Precise **[1 mark]**; all five measurements are within ±0.5 mm of 791.5 mm, which is within the precision of the device (i.e. a steel rule) [accept ±1 mm of 792 mm] **[1 mark]**

Errors and Uncertainties
QUICK TEST (page 14)

1. ±0.5°C (allow ±1°C)
2. the smallest observable difference in a quantity being measured
3. ±0.05 g
4. ±0.1 A
5. the absolute uncertainty
6. $\dfrac{\text{absolute uncertainty of } a}{\text{measured value of } a} \times 100$ or $\dfrac{\Delta a}{a} \times 100$
7. absolute uncertainty $= \pm\left(\dfrac{1}{2} \times \text{range}\right)$
8. 1.22 kΩ
9. ±0.005 kΩ [accept ±0.1 kΩ]
10. ±0.06 kΩ
11. $\dfrac{0.06}{1.22} \times 100 = 4.9 \approx 5\%$

PRACTICE QUESTIONS (page 15)

1. a) $R = \dfrac{V}{I}$ **[1 mark]** $= \dfrac{8.2}{0.8} = 10.25 \approx 10.3\,\Omega$ **[1 mark]**

   b) $V = 8.2 \pm 0.2\,V$ and $I = 0.8 \pm 0.1\,A$

   $\varepsilon V = \dfrac{\Delta V}{V} \times 100$ **[1 mark]** $= \dfrac{0.2}{8.2} \times 100 = 2.4\%$ **[1 mark]**

   $\varepsilon I = \dfrac{\Delta I}{I} \times 100$ **[1 mark]** $= \dfrac{0.1}{0.8} \times 100 = 12.5\%$ **[1 mark]**

   c) (i) $\varepsilon R = \varepsilon V + \varepsilon I$
   $= 2.4\% + 12.5\% = 14.9\%$ **[1 mark]**

   (ii) $\varepsilon R = \dfrac{\Delta R}{R} \rightarrow \Delta R = R \times \varepsilon R$ **[1 mark]**

   $\Delta R = 10.25 \times \dfrac{14.9}{100}$

   $\Delta R = 1.5(3)\,\Omega$ **[1 mark]**

   d) $R = (10.3 \pm 1.5)\,\Omega$ **[1 mark]**

2. a) mean mass $= \dfrac{172.4}{5}$ **[1 mark]** $= 34.48 \approx 34.5\,g$ **[1 mark]**

   b) resolution of scales = 0.1 g
   uncertainty of scales = ±0.05 g **[1 mark]**

   c) uncertainty in measurement
   $= \pm\dfrac{1}{2}(\text{range}) = \pm\dfrac{1}{2}(34.9 - 34.1)$ **[1 mark]**

   $= \pm\dfrac{0.8}{2} = \pm0.4\,g$ **[1 mark]**

   ∴ percentage uncertainty in mass = 1.2% and mass $= (34.5 \pm 0.4)\,g$ **[1 mark]**

d) $l = 2.3 \pm 0.01$ cm
   $V = l^3 = (2.3)^3 = 12.167 \approx 12.2\,\text{cm}^3$ **[1 mark]**

   $\varepsilon V = 3\varepsilon l$ and $\varepsilon l = \dfrac{\Delta l}{l} \times 100 = \dfrac{0.01}{2.3} \times 100$ **[1 mark]**

   $\varepsilon l = 0.43$

   ∴ $\varepsilon V = 3 \times 0.43 = 1.30\%$ **[1 mark]**

   and $\varepsilon V = \dfrac{\Delta V}{V} \times 100 \rightarrow \Delta V = \dfrac{V \times \varepsilon V}{100}$ **[1 mark]**

   $= \dfrac{12.2 \times 1.3}{100}$

   $\Delta V = 0.16\,\text{cm}^3$ **[1 mark]**

e) $\rho = \dfrac{m}{V} = \dfrac{34.5}{12.2} = 2.83\,\text{g cm}^{-3} = 2830\,\text{kg m}^{-3}$
   $\approx 2800\,\text{kg m}^{-3}$ **[1 mark]**

   $\varepsilon \rho = \varepsilon m + \varepsilon V = 1.16 + 1.30 = 2.46\%$ **[1 mark]**

   ∴ $\Delta \rho = \rho \times \dfrac{\varepsilon \rho}{100} = 69.6 \approx 70$ **[1 mark]**

   ∴ $\rho = (2800 \pm 70) \approx (2800 \pm 100)\,\text{kg m}^{-3}$ **[1 mark]**

3. a) $F = kx \rightarrow k = \dfrac{F}{x} = \dfrac{15}{4.6 \times 10^{-3}}$ **[1 mark]** $= 3260.86$

   $k = 3260\,\text{N m}^{-1}$ [accept 3300 N m^{-1}] **[1 mark]**

   b) $\Delta x = \pm0.5\,\text{mm}$, $\Delta F = \pm0.5\,N$

   (i) $\dfrac{\Delta x}{x} = \dfrac{0.5}{4.6} = 0.109$ and $\varepsilon x = 10.9\%$ **[1 mark]**

   (ii) $\dfrac{\Delta F}{F} = \dfrac{0.5}{15} = 0.033$ and $\varepsilon F = 3.3\%$ **[1 mark]**

   c) $\varepsilon k = \varepsilon F + \varepsilon x$
   $= 3.3\% + 10.9\%$
   $= 14.2\%$ **[1 mark]**

   ∴ $\varepsilon k = \dfrac{\Delta k}{k} \times 100 \rightarrow \Delta k = \dfrac{k \times \varepsilon k}{100}$

   $\Delta k = \dfrac{3260 \times 14.2}{100} = 462.9$ **[1 mark]**

   ∴ $k = (3250 \pm 500)\,\text{N m}^{-1}$ **[1 mark]**

   d) 3260, 3300, 3240, 3190, 3140

   mean value $= \dfrac{16130}{5} = 3226\,\text{Nm}^{-1}$
   $\approx 3230\,\text{Nm}^{-1}$ **[1 mark]**

   uncertainty in spring constant $= \pm\dfrac{1}{2}(\text{range})$

   $= \pm\dfrac{1}{2}(3300 - 3140)$

   $= \pm80\,\text{N m}^{-1}$ **[1 mark]**

   ∴ $k = (3230 \pm 80)\,\text{Nm}^{-1}$ **[1 mark]**

Graphs
QUICK TEST (page 18)

1. the y-axis
2. m and c, respectively
3. **(i)** straight line **(ii)** parabolic
   (iii) reciprocal (accept inversely proportional)
4. **(i)** V and I **(ii)** E and v^2 **(iii)** E and $\frac{1}{\lambda}$
5. **(i)** R **(ii)** $\frac{1}{2}m$ **(iii)** hc

6. parabolic
7. c^2
8. resistivity × length; Ωm^2
9. energy
10. velocity

PRACTICE QUESTIONS (page 19)

1. a)

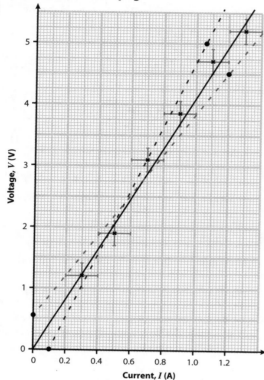

[4 marks: 1 mark for axes labelled correctly; 1 mark for points plotted correctly; 1 mark for error bars drawn; 1 mark for line of best fit drawn correctly]

b) gradient (line of best fit) $m = \dfrac{\Delta y}{\Delta x} = \dfrac{5-0}{1.22-0}$ [1 mark]

$= 4.1\,V\,A^{-1}$ [1 mark]

c) shallowest/steepest lines drawn through error bars

shallowest gradient $= \dfrac{4.5-0.55}{1.2-0} = 3.3\,V\,A^{-1}$ [1 mark]

steepest gradient $= \dfrac{5.0-0}{1.07-0.1} = 5.2\,V\,A^{-1}$ [1 mark]

d) gradient $m = (4.1 \pm 1.1)\,V\,A^{-1}$ [2 marks]; allow $m = (4 \pm 1)$ [Accept Ω in place of $V\,A^{-1}$]

2. a)

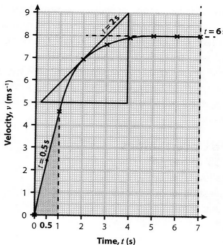

[4 marks: 1 mark for correct values used on x and y-axes; 1 mark for labelling the x and y-axes; 1 mark for plotting points correctly; 1 mark for drawing a smooth lines through their points]

b) (i) $t = 0.5\,s$ $m = \dfrac{4.2-0}{0.9-0} = 4.7\,m\,s^{-2}$ [1 mark]

(ii) $t = 2\,s$ $m = \dfrac{9-5}{3.9-0.2} = 1.1\,m\,s^{-2}$ [1 mark]

(iii) $t = 6\,s$ $m = 0\,m\,s^{-2}$ [1 mark]

c) Gradient represents acceleration [1 mark], with units of $m\,s^{-2}$ [1 mark]

d) (i) area under curve, $0-1\,s = \dfrac{1}{2}(1)(4.6) \approx 2.3\,m$ [1 mark]

(ii) area under curve, $0-7\,s = 47$ full squares [1 mark]
$\approx 47 \times 1 = 47\,m$ [1 mark]

e) Area under curve represents distance or displacement [1 mark]

3. a)

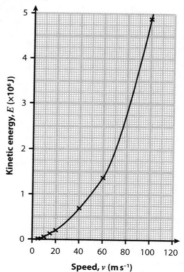

[4 marks: 1 mark for correct axes (Energy × 10^6 J); 1 mark for axes correctly labelled; 1 mark for points plotted correctly; 1 mark for smooth curve drawn through points]

b) parabolic graph, of the form $y = kx^2$ **[1 mark]**

hence $E = \frac{1}{2}mv^2$, the equation for kinetic energy **[1 mark]**

c)

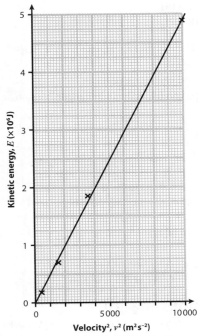

[2 marks: 1 mark for correct axes on redrawn graph; 1 mark for points plotted and line drawn correctly]

gradient $m = \dfrac{4.9 \times 10^6 - 0}{10\,000 - 0} = 490\ \text{J m}^{-2}\,\text{s}^2$ **[1 mark]**

d) Gradient represents $\frac{1}{2}$ mass $\approx \frac{1}{2}(1000) = 500\ \text{kg}$ **[1 mark]**

Day 2

Nuclei and Particle Delay
QUICK TEST (page 22)

1. 11 protons, 13 neutrons
2. 92 protons, 92 electrons and 143 neutrons
3. atoms with the same number of protons but different number of neutrons
4. $^{63}_{29}\text{Cu}$
5. $^{65}_{29}\text{Cu}$
6. neutrons
7. $+4.25 \times 10^7\ \text{C kg}^{-1}$
8. $+1.37 \times 10^7\ \text{C kg}^{-1}$
9. $+2e = 3.2 \times 10^{-19}\ \text{C}$; $3.04 \times 10^6\ \text{C kg}^{-1}$
10. strong nuclear force
11. electromagnetic or electrostatic force
12. $^{229}_{90}\text{Th} \rightarrow {}^{225}_{88}\text{Ra} + {}^{4}_{2}\alpha$
13. $^{14}_{6}\text{C} \rightarrow {}^{14}_{7}\text{N} + {}^{0}_{-1}\beta + \bar{\nu}$

PRACTICE QUESTIONS (page 23)

1. a) Specific charge is the ratio of total charge to total mass $\left(\dfrac{Q}{m}\right)$ for the nucleus; it has units of coulomb per kg (C kg^{-1}) **[1 mark]**

b) $Q = 1.60 \times 10^{-19}\ \text{C}$ $m = 1.67 \times 10^{-27}\ \text{kg}$

specific charge $= \dfrac{Q}{m} = \dfrac{1.60 \times 10^{-19}}{1.67 \times 10^{-27}}$ **[1 mark]**

$\qquad\qquad = 9.6 \times 10^5\ \text{C kg}^{-1}$ **[1 mark]**

c) (i) neutron (as $Q = 0$) **[1 mark]**
    (ii) proton (as m is larger) **[1 mark]**

d) $^{24}_{11}\text{Mg}$ ionised:
    (i) 11 protons, 13 neutrons and 10 electrons **[2 marks]**
    (ii) $Q = +1e = 1.60 \times 10^{-19}\ \text{C}$ **[1 mark]**
    (iii) specific charge

$= \dfrac{Q}{m} = \dfrac{1.60 \times 10^{-19}}{\left(24 \times 1.67 \times 10^{-27}\right) + \left(10 \times 9.11 \times 10^{-31}\right)}$ **[1 mark]**

$= 4.0 \times 10^6\ \text{C kg}^{-1}$ **[1 mark]**

2. a) antineutrino (or electron antineutrino) **[1 mark]**

b) $A = 65$ **[1 mark]** and $Z = 29$ **[1 mark]**

c) electron **[1 mark]**

d) Nickel has too many neutrons; changes neutrons into protons **[1 mark]**

e) The antineutrino carries away the excess kinetic energy **[1 mark]** so that $E_k^\beta + E_k^\nu$ is constant **[1 mark]**

3. a) The strong nuclear force acts between protons; it is a repulsive force below a separation distance of less than 0.4 fm **[1 mark]** but attractive between 0.4 fm and about 3 fm **[1 mark]**; beyond 3 fm it rapidly decreases to zero **[1 mark]**

Forces between protons

[2 marks]

b) Heavy nuclei have too many protons for the strong nuclear force to keep them stable **[1 mark]**; to make themselves more stable they emit alpha particles, i.e. helium nuclei **[1 mark]** via the interaction $^{A}_{Z}\text{X} \rightarrow {}^{A-4}_{Z-2}\text{Y} + {}^{4}_{2}\text{He}$ so that it reduces by 2 protons and 2 neutrons after each decay mode **[2 marks]**

c) $^{238}_{92}U \rightarrow {}^{234}_{90}Th + {}^{4}_{2}He$ **[2 marks: 1 mark for each correct nuclear term]**

Particles, Antiparticles, Exchange Particles and Quarks
QUICK TEST (page 27)
1. $2 \times 0.511\,\text{MeV} = 1.022\,\text{MeV}$
2. The excess is shared between the photons which are now more energetic.
3. $2 \times 938 = 1876\,\text{MeV}$
4. strong nuclear force
5. up, down, strange, anti-up, anti-down and anti-strange quarks
6. mesons
7. pions and kaons
8. kaons
9. kaons
10. three types: π^{+}, π^{-} and π^{0}
11. (udd) and (u$\bar{s}$)
12. two possible quark structures: (d$\bar{s}$) and ($\bar{d}$s)
13. exchange particles (or gauge/virtual bosons)

PRACTICE QUESTIONS (page 27)
1. a) Hadrons are a group of (sub-atomic) particles that interact via the strong nuclear force **[1 mark]**
   b) three quarks, i.e. (qqq) **[1 mark]**; e.g. proton (uud) or neutron (udd) **[1 mark]**
   c) Mesons are made from quark–antiquark pairs (q$\bar{q}$) **[1 mark]**
   d) pions and kaons **[1 mark]**; pions contain u, d, $\bar{u}$ and/or $\bar{d}$ quarks **[1 mark]**; kaons also contain s or $\bar{s}$ quarks **[1 mark]**
2. a) antibaryons **[1 mark]**
   b) antiproton ($\bar{u}\bar{u}\bar{d}$)
      or
      or antineutron ($\bar{u}\bar{d}\bar{d}$) **[2 marks]**
   c) antiproton charge $-1 = -\frac{2}{3} - \frac{2}{3} + \frac{1}{3}$
      or
      antineutron charge $0 = -\frac{2}{3} + \frac{1}{3} + \frac{1}{3}$ **[2 marks]**
   d) The antiproton or antineutron annihilates **[1 mark]** with a proton or neutron, respectively, to create high-energy photons **[1 mark]**
3. a) (i) A quark is a fundamental particle **[1 mark]** that makes up particles such as protons and neutrons; quarks exert the strong nuclear force on each other **[1 mark]**
      (ii) Quark confinement means that 'free' quarks are never observed **[1 mark]**
   b) gluon **[1 mark]**; has an extremely short range of action of $\sim 10^{-15}\,\text{m}$ **[1 mark]**
   c) electron, up quark, down quark, proton, neutron **[1 mark]**

d) difference: different quark structure **[1 mark]**
   similarity: same charge **or** composed of three quarks **[1 mark]**
e) (i) $\left(\bar{u}\bar{d}\bar{d}\right)$ **[1 mark]**
   (ii) 0 **[1 mark]**

The Standard Model and Conservation Rules
QUICK TEST (page 30)
1. leptons, quarks and exchange particles
2. (i) hadron and baryon (ii) lepton (iii) hadron, meson and kaon
3. (i) 0 (ii) -1 (iii) $+\frac{1}{3}$
4. (i) 0 (ii) $+1$ (iii) -1
5. Yes. Charge, baryon number, lepton number (and strangeness) are all conserved
6. $0 \rightarrow 0 + 1 - 1$; so the decay mode is possible
7. (sss)
8. (i) weak interaction (ii) no
9. (i) charge: $+1 + 1 \rightarrow +1 + 0 + 1$;
   (ii) baryon number: $+1 + 1 \rightarrow +1 + 1 + 0$
10. Baryon number is not conserved: $+1 - 1 \rightarrow +1 + 0$

PRACTICE QUESTIONS (page 31)
1. a) similarity: electrically neutral or no charge **[1 mark]**
      difference: neutron is a hadron/baryon; ν_e is a lepton or ν_e is a fundamental particle **[1 mark]**
   b) electromagnetic – photon **[1 mark]**
      gravitational – graviton **[1 mark]**
      weak – $W^{\pm}$ / Z boson **[1 mark]**
      strong – gluon **[1 mark]**
   c) weak interaction **[1 mark]**
   d) $n + \nu_e \rightarrow p + e^{-}$ **[2 marks]**
2. a) Hadrons are particles that are made up from quarks **[1 mark]** and are therefore subject to the strong nuclear force **[1 mark]**
   b) baryons and mesons **[1 mark]**
      baryons (qqq) or ($\bar{q}\bar{q}\bar{q}$); mesons (q$\bar{q}$) **[1 mark]**
   c) neutron (udd) **[1 mark]**; pion: π^{+} (u$\bar{d}$) or kaon: K^{+} (u$\bar{s}$) **[1 mark]**
3. a) (uds) **[1 mark]**
   b)

	$\Lambda^{\circ} \rightarrow p + \pi$	
charge, Q:	$0 \rightarrow 1 + -1$	✓ **[1 mark]**
baryon number, B:	$1 \rightarrow 1 + 0$	✓ **[1 mark]**
lepton number, L:	$0 \rightarrow 0 + 0$	✓ **[1 mark]**
strangeness, S:	$-1 \rightarrow 0 + 0$	✗ **[1 mark]**

   c) The decay process involves the weak interaction **[1 mark]**
   d) $\Lambda^{0} \rightarrow n + \pi^{0}$, i.e. X is π^{0} **[1 mark]**; π^{0}: (u$\bar{u}$) or (d$\bar{d}$) **[1 mark]**
   e) difference: π^{0} is neutral but π^{-} is negatively charged **[1 mark]**
      similarity: both are pions **[1 mark]**

Particle interactions and Feynman Diagrams

QUICK TEST (page 35)

1. a pictorial representation of a particle interaction
2. by straight lines with arrows
3. by wavy lines/spiral curves
4. by the vertices between lines
5. because their lifetime is so short that they exist only for an extremely short time and therefore are not directly observed

6.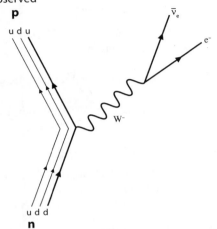

7. W^+ or W^-
8. from left to right unless indicated otherwise
9. A gluon is the exchange particle between quarks showing the strong interaction; shown by a spiral curve
10. weak interaction
11. electron antineutrino and muon neutrino, or electron neutrino and muon antineutrino
12. photon
13. (i) $\mu^- \rightarrow e^- + \bar{\nu}_e + \nu_\mu$

(ii)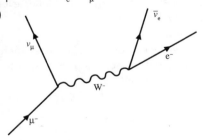

PRACTICE QUESTIONS (page 35)

1. a)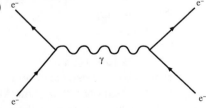

[3 marks]

b) electromagnetic force **[1 mark]**; photon **[1 mark]**

c)
$$e^- + e^- \rightarrow e^- + e^-$$

charge, Q:	$-1 - 1 \rightarrow -1 - 1$	✓ **[1 mark]**
lepton number, L:	$+1 + 1 \rightarrow +1 + 1$	✓ **[1 mark]**
baryon number, B:	$0 + 0 \rightarrow 0 + 0$	✓ **[1 mark]**

2. a) $p \rightarrow n + e^+ + \nu_e$ **[2 marks]**

b)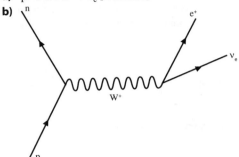

[3 marks]

c) weak interaction **[1 mark]**; W^+ boson **[1 mark]**

d) proton (uud) changes into a neutron (udd) **[1 mark]** or an up quark changes into a down quark

e)
$$p \rightarrow n + e^+ + \nu_e$$

charge, Q:	$+1 \rightarrow 0 + 1 + 0$	✓ **[1 mark]**
lepton number, L:	$0 \rightarrow 0 - 1 + 1$	✓ **[1 mark]**
baryon number, B:	$+1 \rightarrow +1 + 0 + 0$	✓ **[1 mark]**

3. a)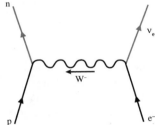

[3 marks: 1 mark for each correct label of electron, electron neutrino and $W^{(-)}$ boson]

b) weak interaction **[1 mark]**

c) $p + e^- \rightarrow n + \nu_e$ **[2 marks]**

d)
$$p + e^- \rightarrow n + \nu_e$$

charge, Q:	$+1 - 1 \rightarrow 0 + 0$	✓ **[1 mark]**
lepton number, L:	$0 + 1 \rightarrow 0 + 1$	✓ **[1 mark]**
baryon number, B:	$+1 + 0 \rightarrow +1 + 0$	✓ **[1 mark]**

Day 3

Photons and the Quantum of Light

QUICK TEST (page 38)

1. photoelectrons
2. the emission of 'free' electrons near the surface of a metal by the absorption of visible or ultraviolet radiation that break the bonds holding the electrons
3. the frequency of the incident radiation below which no photoelectrons are emitted
4. $f_0 = \frac{\Phi}{h}$

5. $hf = \Phi + E_k^{max}$
6. a photon
7. $E_k^{max} = \frac{1}{2}mv_{max}^2 = hf - \Phi$
8. multiply by 1.6×10^{-19}
9. visible and UV light
10. The work function depends on the minimum amount of energy required to break the bonds that bind the electrons in that particular metal
11. 3.1×10^{-19} J
12. 5.6×10^{14} Hz
13. the potential well
14. 0.8 eV

PRACTICE QUESTIONS (page 39)

1. $\Phi = 4.64 \times 10^{-19}$ J
   $\lambda = 0.4\,\mu m$

   a) $c = f\lambda \rightarrow f = \dfrac{c}{\lambda} = \dfrac{3 \times 10^8}{0.4 \times 10^{-6}} = 7.5 \times 10^{14}$ Hz **[1 mark]**

   $E = hf = 6.63 \times 10^{-34} \times 7.5 \times 10^{14}$
   $= 4.97 \times 10^{-19}$ J $\approx 5.0 \times 10^{-19}$ J **[1 mark]**
   $= \dfrac{4.97 \times 10^{-19}}{1.6 \times 10^{-19}} = 3.11$ eV ≈ 3.1 eV **[1 mark]**

   b) $\Phi = \dfrac{4.64 \times 10^{-19}}{1.6 \times 10^{-19}} = 2.9$ eV **[1 mark]**

   c) $E = hf = \Phi + \frac{1}{2}mv_{max}^2$
   $\frac{1}{2}mv_{max}^2 = E - \Phi$ **[1 mark]**
   $= 3.11 - 2.9$
   $= 0.21$ eV **[1 mark]**

   d) λ is smaller $\therefore f$ is higher $\therefore E$ is larger **[1 mark]**
   If E is larger, $\frac{1}{2}mv_{max}^2$ is also larger (0.65 eV) **[1 mark]**

2. a) The incident photon energy is given by $E = hf$ **[1 mark]**.
   The threshold energy for an electron to escape from the metal (i.e. the work function) is given by $\Phi = hf_0$ **[1 mark]**; f_0 is the threshold frequency
   If $E < \Phi$, or $f < f_0$, then no photoelectrons are emitted **[1 mark]**

   b) The wave theory of light suggests that eventually the electrons will accumulate or absorb sufficient energy from the incident radiation to be emitted from the surface of a metal **[1 mark]**
   This is not observed and emission is independent of the intensity of the incident radiation **[1 mark]**

   c) $\Phi = 7.2 \times 10^{-19}$ J $\qquad \Phi = hf_0$
   $\therefore f_0 = \dfrac{\Phi}{h} = \dfrac{7.2 \times 10^{-19}}{6.6 \times 10^{-34}}$ **[1 mark]** $= 1.09 \times 10^{15}$ Hz
   $\approx 1.1 \times 10^{15}$ Hz **[1 mark]**

   d) $E = \Phi + \frac{1}{2}mv_{max}^2$
   $E = 7.2 \times 10^{-19} + 3.4 \times 10^{-19}$
   $E = 10.6 \times 10^{-19}$ J **[1 mark]**

$E = \dfrac{10.6 \times 10^{-19}}{1.6 \times 10^{-19}}$ eV
$E = 6.63$ eV ≈ 6.6 eV **[1 mark]**

$E = hf \rightarrow f = \dfrac{E}{h} = \dfrac{10.6 \times 10^{-19}}{6.63 \times 10^{-34}}$ Hz
$f = 1.6 \times 10^{15}$ Hz **[1 mark]**

3. a) gradient represents Planck's constant **[1 mark]**

   b) $m = \dfrac{\Delta y}{\Delta x} = \dfrac{(3.68 - 0.5) \times 10^{-19}}{(10 - 5) \times 10^{14}}$ **[1 mark]**
   $= 6.36 \times 10^{-34}$ Js $\approx 6.4 \times 10^{-34}$ Js **[1 mark]**

   c) x-intercept represents the threshold frequency **[1 mark]**

   d) y-intercept is the work function or threshold energy **[1 mark]**

   e) Crossing point on x-axis is 4.25×10^{14} Hz **[1 mark]**
   $\Phi = hf_0 = 2.82 \times 10^{-19}$ J $\approx 2.8 \times 10^{-19}$ J **[1 mark]**

   f) $\lambda = 0.42\,\mu m \quad E = \dfrac{hc}{\lambda} = \dfrac{6.63 \times 10^{-34} \times 3 \times 10^8}{0.42 \times 10^{-6}}$ **[1 mark]**
   $E = 4.74 \times 10^{-19}$ J $\approx 4.7 \times 10^{-19}$ J **[1 mark]**
   $\frac{1}{2}mv_{max}^2 = E - \Phi = 4.74 \times 10^{-19} - 2.82 \times 10^{-19}$
   $= 1.92 \times 10^{-19}$ J $\approx 1.9 \times 10^{-19}$ J **[1 mark]**

Energy Levels and Transitions
QUICK TEST (page 42)

1. the minimum energy required to remove an electron from an atom
2. ground state
3. when electrons absorb exactly the right amount of energy to move to higher energy levels
4. The excess energy is emitted in the form of a photon
5. $n = \infty$
6. the difference between the energy of the higher level and that of the lower level
7. the visible spectrum
8. It moves from the lower energy level to a higher energy level corresponding to the exact energy of the photon
9. $\Delta E = -1.51 - (-3.40) = 1.89$ eV;
   $\Delta E = 1.89 \times 1.60 \times 10^{-19} = 3.02 \times 10^{-19}$ J;
   $\lambda = \dfrac{hc}{\Delta E} = 6.59 \times 10^{-7}$ m $= 659$ nm
10. red part of the spectrum
11. photon absorption or electron collision

PRACTICE QUESTIONS (page 43)

1. a) (i) $E = 12.5$ eV **[1 mark]**
      (ii) $E = 12.5 \times 1.6 \times 10^{-19}$
      $= 20 \times 10^{-19}$ J or 2.0×10^{-18} J **[1 mark]**

   b) The incoming electron allows 12.09 eV to be absorbed by an atomic electron in the ground state **[1 mark]**; this atomic electron is excited to level $n = 3$; the free electron leaves the atom with only 0.41 eV of kinetic energy **[1 mark]**

   c) $3 \rightarrow 2$ then $2 \rightarrow 1$ or $3 \rightarrow 1$ **[1 mark]**

d) 3→2: $\Delta E = E_3 - E_2 = -1.51 - (-3.4)$
$\quad\quad\quad\quad \Delta E = 1.9\,\text{eV}$ **[1 mark]**

$\quad\quad$ 3→1: $\Delta E = E_3 - E_1 = -1.51 - (-13.6)$
$\quad\quad\quad\quad \Delta E = 12.1\,\text{eV}$ **[1 mark]**

$\quad\quad$ 2→1: $\Delta E = E_2 - E_1 = -3.4 - (-13.6)$
$\quad\quad\quad\quad \Delta E = 10.2\,\text{eV}$ **[1 mark]**

2. a) $E = -122.4 \times 1.6 \times 10^{-1}\,\text{J}$
$\quad\quad E = -1.96 \times 10^{-17}\,\text{J} \approx -2.0 \times 10^{-17}\,\text{J}$ **[1 mark]**

b) ionisation energy is 122.4 eV **[1 mark]**

c) (i) 3→2 then 2→1, or 3→1 **[1 mark]**

(ii) 3→2: $\Delta E = E_3 - E_2 = -13.6 - (-30.6)$
$\quad\quad\quad\quad \Delta E = 17.0\,\text{eV}$ **[1 mark]**

$\quad\quad$ 3→1: $\Delta E = E_3 - E_1 = -13.6 - (-122.4)$
$\quad\quad\quad\quad \Delta E = 109\,\text{eV}$ **[1 mark]**

$\quad\quad$ 2→1: $\Delta E = E_2 - E_1 = -30.6 - (-122.4)$
$\quad\quad\quad\quad \Delta E = 91.8\,\text{eV}$ **[1 mark]**

d) 5→4
$\Delta E = E_5 - E_4 = -4.9 - (-7.65)$
$\Delta E = 2.75\,\text{eV}$ **[1 mark]**

$E = \dfrac{hc}{\lambda} \rightarrow \lambda = \dfrac{hc}{E}$

$\lambda = \dfrac{6.63 \times 10^{-34} \times 3 \times 10^8}{\left(2.75 \times 1.6 \times 10^{-19}\right)} = 4.52 \times 10^{-7}\,\text{m}$

$\lambda = 452\,\text{nm}$ **[1 mark]**
visible region **[1 mark]**

Emission and Absorption Spectra
QUICK TEST (page 46)

1. a series of bright lines on a dark background due to electrons falling from higher energy levels to lower energy levels and emitting photons of a particular energy (frequency) that is characteristic of the gas
2. continuous spectrum
3. prism, diffraction grating
4. when gas atoms absorb photons from white light, corresponding to discrete energy levels in the gas
5. They are characteristic of the discrete energy levels within the gas
6. a mixture of electrons and ions in a gas
7. thermionic emission
8. mercury vapour and a phosphor coating
9. The photons are of the correct energy to be absorbed by the phosphor coating in order to excite electrons in the coating; these electrons in turn fall back to the ground state, emitting photons in the visible region
10. astronomy/stellar structure

PRACTICE QUESTIONS (page 46)
1. a) When the tube is switched on, the cathode filament is heated causing electrons to be released by thermionic emission **[1 mark]**. A potential difference across the tube accelerates these 'free' electrons **[1 mark]**. Collisions with electrons in the mercury atoms produces excitation and ionisation **[1 mark]**. When

electrons in the mercury atoms return to their ground state, they release photons in the UV range **[1 mark]**.

b) Fluorescence occurs when a material such as phosphor absorbs UV radiation (short wavelength) **[1 mark]** and emits radiation in the visible region (longer wavelength) **[1 mark]**

c) The phosphor coating absorbs the UV radiation **[1 mark]**, which excites the electrons in the phosphor **[1 mark]**. De-excitation of these electrons to the ground state produces photon emission in the visible region **[1 mark]** **[Deduct 1 mark from each answer if the quality of the written communication is poor]**

2. a) An emission spectrum is a series of distinct, coloured lines on a dark background **[1 mark]** corresponding to the emission of photons when excited electrons return to a lower energy level **[1 mark]**

b) the lowest energy state an electron can be in **[1 mark]**

c) (i) $\Delta E = 1.86\,\text{eV}$ $\quad \therefore \lambda = \dfrac{hc}{\Delta E} = \dfrac{6.63 \times 10^{-34} \times 3 \times 10^8}{\left(1.86 \times 1.6 \times 10^{-19}\right)}$
$\quad\quad \lambda = 668\,\text{nm}$, red **[1 mark]**

(ii) $\Delta E = 2.79\,\text{eV}$ $\quad \lambda = 446\,\text{nm}$, blue **[1 mark]**

d) $\Delta E = -4.4 - (-24.6)\,\text{eV}$
$\quad \Delta E = 20.2\,\text{eV}$ **[1 mark]**

$\lambda = \dfrac{hc}{\Delta E} = \dfrac{6.63 \times 10^{-34} \times 3 \times 10^8}{\left(20.2 \times 1.6 \times 10^{-19}\right)}$
$\lambda = 61.5\,\text{nm} \approx 62\,\text{nm}$ **[1 mark]**
in extreme UV region **[1 mark]**

3. a) An absorption spectrum is produced when light from a hot body (e.g. the Sun's interior/core) passes through a cooler gas (e.g. the Sun's atmosphere) **[1 mark]**; dark lines represent photon absorption **[1 mark]**; subsequent de-excitation is in all directions and hence lines remain dark; the dark lines appear on a continuous spectrum background **[1 mark]**

b) Transition $n = 3$ to $n = 2$
$\Delta E = -1.51 - (-3.40) = 1.89\,\text{eV}$ **[1 mark]**

$\lambda = \dfrac{hc}{\Delta E} = \dfrac{6.63 \times 10^{-34} \times 3 \times 10^8}{\left(1.89 \times 1.6 \times 10^{-19}\right)}$
$\lambda = 658\,\text{nm}$ **[1 mark]**
This represents the H_α transition **[1 mark]**

c) in the red part of the visible spectrum **[1 mark]**

Wave–Particle Duality
QUICK TEST (page 50)

1. diffraction, interference
2. photoelectric effect
3. the corresponding wavelength of a moving particle
4. its mass and velocity, i.e. its momentum
5. shorter wavelength
6. reduced energy
7. the idea that matter and radiation can be described sometimes as a wave and sometimes as a particle
8. $E = \dfrac{hc}{\lambda}$

9. It showed that light waves behaved like particles with an energy given by $E = hf$

10. The spacing between the lines is reduced

11. electron diffraction in crystals/electron microscopy

12. all three

PRACTICE QUESTIONS (page 50)

1. **a)** Wave–particle duality is the notion that matter and radiation can be described sometimes using a wave model **[1 mark]** and sometimes using a particle model **[1 mark]**

   b) $v = 5.6 \times 10^5 \, \text{m s}^{-1}$, $m_e = 9.11 \times 10^{-31} \, \text{kg}$

   $$\lambda = \frac{h}{mv} = \frac{6.63 \times 10^{-34}}{\left(5.6 \times 10^5\right)\left(9.11 \times 10^{-31}\right)} \text{ [1 mark]}$$

   $= 1.3 \times 10^{-9} \, \text{m} \, (1.3 \, \text{nm})$ **[1 mark]**

   c) $m_p = 1.67 \times 10^{-27} \, \text{kg}$

   $$v = \frac{h}{m\lambda} = \frac{6.63 \times 10^{-34}}{\left(1.67 \times 10^{-27}\right)\left(1.3 \times 10^{-9}\right)} \text{ [1 mark]}$$

   $= 305.4 \approx 300 \, \text{m s}^{-1}$ **[1 mark]**

   d) Electrons would be better suited to study the structure of materials because:

   (i) small size/mass; less destructive **[1 mark]**

   (ii) easily obtained via thermionic emission **[1 mark]**

2. **a)** $m_p = 1.67 \times 10^{-27} \, \text{kg}$ $v = c = 3 \times 10^8 \, \text{m s}^{-1}$

   $$\lambda = \frac{h}{mc} = \frac{6.63 \times 10^{-34}}{1.67 \times 10^{-27} \times 3 \times 10^8} \text{ [1 mark]}$$

   $= 1.32 \times 10^{-15} \, \text{m}$ **[1 mark]**

   b) $1.32 \times 10^{-15} \, \text{m} = 1.32 \, \text{fm}$

   c) Structures that are much smaller than the nucleus **[1 mark]**; wavelengths significantly smaller than the size of the nucleus **[1 mark]**

   d) Wavelength of proton is much less than the size of the nuclear structure **[1 mark]**; diffraction effects will not be significant **[1 mark]**

3. **a)** $E = 100 \, \text{keV}$

   $E = 100 \times 10^3 \times 1.60 \times 10^{-19} \, \text{J}$ **[1 mark]**

   $E = 1.60 \times 10^{-14} \, \text{J}$ **[1 mark]**

   b) $E_k = \frac{1}{2} mv^2$

   $$v^2 = \frac{2E_k}{m} \Rightarrow v = \sqrt{\frac{2E_k}{m}} \text{ [1 mark]}$$

   $$v = \sqrt{\frac{2 \times 1.60 \times 10^{-14}}{9.11 \times 10^{-31}}} \text{ [1 mark]} = 1.87 \times 10^8 \, \text{m s}^{-1}$$

   $v \approx 1.9 \times 10^8 \, \text{m s}^{-1}$ **[1 mark]**

   c) $\lambda = \frac{h}{mv} = \frac{6.63 \times 10^{-34}}{9.11 \times 10^{-31} \times 1.9 \times 10^8}$ **[1 mark]**

   $\lambda = 3.83 \times 10^{-12} \, \text{m}$

   $\lambda = 3.83 \times 10^{-3} \, \text{nm} \approx 3.8 \times 10^{-3} \, \text{nm}$ **[1 mark]**

   d) $\lambda_{\text{light}} \approx 500 \, \text{nm}$

   $\text{ratio} = \frac{3.83 \times 10^{-3}}{500} = 7.66 \times 10^{-6} \approx 8 \times 10^{-6}$ **[1 mark]**

Day 4

Waves and Vibrations

QUICK TEST (page 54)

1. s^{-1} or Hz (hertz)

2. the maximum displacement from the equilibrium position

3. degrees or radians

4. $c = f\lambda$

5. $f = \frac{1}{T}$

6. The vibrations of a longitudinal wave are parallel to the direction of travel

7. unpolarised

8. [any two from] electromagnetic waves (such as X-rays, microwaves, IR or UV), secondary seismic waves, waves on a string, water ripples

9. It becomes partially polarised

10. Broadcast TV signals are polarised, so receivers also have to be orientated to receive these polarised signals

11. the plane in which a transverse wave vibrates

12. The vibrations of the wave occur in all planes.

13. $340 \, \text{m s}^{-1}$

14. $2.5 \times 10^{-4} \, \text{s}$ or $0.25 \, \text{ms}$

15. $650 \, \text{nm}$ or $650 \times 10^{-9} \, \text{m}$ or $6.5 \times 10^{-7} \, \text{m}$

PRACTICE QUESTIONS (page 55)

1. **a)** A longitudinal wave is a primary seismic wave or P-wave **[1 mark]**

   A transverse wave is a secondary seismic wave or S-wave **[1 mark]**

   b) phase difference: A: 90° or $\frac{\pi}{2}$ radians **[1 mark]**

   B: 540° or 3π radians **[1 mark]**

   C: 720° or 4π radians **[1 mark]**

   c) $f = 0.60 \, \text{Hz}$ $v_t = 4.8 \, \text{km s}^{-1}$

   (i) $v_t = f\lambda \rightarrow \lambda = \frac{v_t}{f} = \frac{4.8 \times 10^{-3}}{0.60}$ **[1 mark]** $= 8000 \, \text{m}$

   $\lambda = 8 \, \text{km}$ **[1 mark]**

   (ii) $T = \frac{1}{f} = \frac{1}{0.60} = 1.67 \, \text{s}$ **[1 mark]**

   d) $v_l = 7.8 \, \text{km s}^{-1}$ $v_t = 4.8 \, \text{km s}^{-1}$

   $\Delta v = 3.0 \, \text{km s}^{-1}$ $\Delta t = 100 \, \text{s}$

   $\therefore d = \Delta v \Delta t = 3.0 \times 100 = 300 \, \text{km}$ **[2 marks]**

2. **a)** an electromagnetic wave or EM wave or transverse wave **[1 mark]**

   b) Attach one end **[1 mark]**; provide an up-and-down movement to the other end, i.e. vertical oscillation **[1 mark]**, to mimic a transverse, oscillating wave

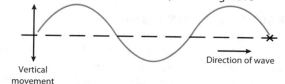

Vertical movement

Direction of wave

[1 mark for correct diagram]

c) Microwaves (any transverse waves) undergo a phase change of 180° (π radians) **[1 mark]**

d) $f = 2.45\,\text{GHz}$ $\qquad c = f\lambda \rightarrow \lambda = \dfrac{c}{f}$

$\lambda = \dfrac{3.0 \times 10^8}{2.45 \times 10^9} = 0.1224\,\text{m} \approx 0.122\,\text{m or } 12.2\,\text{cm}$ **[2 marks]**

e) Microwaves are used to transmit mobile phone signals or in communication **[1 mark]**

3. a) A longitudinal wave vibrates parallel to the direction of motion or to the direction that energy travels in **[1 mark]**

b) **Any two from:** sound waves; primary seismic waves; ultrasound; pulses on a Slinky **[2 marks]**

c) They are all mechanical waves **[1 mark]**; longitudinal waves need a medium to travel through **[1 mark]**

d) Wavelength of a longitudinal wave is the distance between successive compressions or successive rarefactions **[1 mark]**

e) $\lambda_{\text{air}} = \dfrac{v_{\text{air}}}{f} = \dfrac{330}{0.1 \times 10^6} = 3.3 \times 10^{-3}\,\text{m} = 3.3\,\text{mm}$ **[1 mark]**

$\lambda_{\text{water}} = \dfrac{v_{\text{water}}}{f} = \dfrac{1400}{0.1 \times 10^6} = 0.014\,\text{m} = 14\,\text{mm}$ **[1 mark]**

Stationary Waves
QUICK TEST (page 58)

1. from the superposition of two progressive waves with the same frequency (wavelength) and the same amplitude moving in opposite directions

2. nodes and antinodes, respectively

3. The two sources must have the same frequency (wavelength) and a fixed or constant phase difference

4. $\dfrac{1080}{360} = 3$, i.e. an integer multiple, and therefore in phase

5. When two waves cross, the resultant displacement is the (vector) sum of their individual displacements

6. constructive interference (reinforcement)

7. $\frac{1}{2}\lambda = 0.8$, so $\lambda = 1.6\,\text{m}$

PRACTICE QUESTIONS (page 59)

1. a) A stationary or standing wave is a wave formed by the superposition of two progressive waves of the same frequency *and* amplitude **[1 mark]** travelling in opposite directions **[1 mark]**

b) $f = 2.45\,\text{GHz} = 2.45 \times 10^9\,\text{Hz}$

$c = f\lambda \rightarrow \lambda = \dfrac{c}{f} = \dfrac{3.0 \times 10^8}{2.45 \times 10^9}$ **[1 mark]** $= 0.122\,\text{m}$

$\lambda = 12.2\,\text{cm} \approx 12\,\text{cm}$ **[1 mark]**

c)
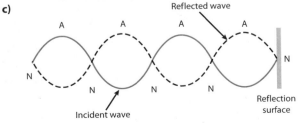
Reflected wave
Incident wave
Reflection surface

[2 marks: 1 mark for correct incident/reflected wave, 1 mark for correct labelling of nodes/antinodes]

d) $\lambda = 12.2\,\text{cm}$

∴ distance between nodes $= \dfrac{\lambda}{2} = 6.1\,\text{cm} \approx 6.0\,\text{cm}$ **[1 mark]**

e) Holes in the chocolate would appear at the antinode positions **[1 mark]**, i.e. separated by a distance of $\sim 6\,\text{cm}$ **[1 mark]**

2. a)
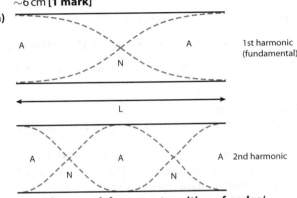
1st harmonic (fundamental)

2nd harmonic

[3 marks: 1 mark for correct position of nodes/antinodes; 1 mark (each) for correct waves in both diagrams]

b) 1st harmonic: $L = \dfrac{\lambda}{2} \rightarrow \lambda = 2L = 2.4\,\text{m}$

$f_1 = \dfrac{v}{\lambda} = \dfrac{330}{2.4} = 137.5\,\text{Hz} \approx 138\,\text{Hz}$ **[1 mark]**

2nd harmonic: $L = \lambda \rightarrow \lambda = 1.2\,\text{m}$

$f_2 = \dfrac{v}{\lambda_1} = \dfrac{330}{1.2} = 275\,\text{Hz}$ **[1 mark]** $\left(\equiv 2f_0\right)$

3rd harmonic: $L = \dfrac{3\lambda}{2} \rightarrow \lambda = \dfrac{2L}{3} = 0.8\,\text{m}$

$f_3 = \dfrac{330}{0.8} = 412.5\,\text{Hz}$

$\approx 413\,\text{Hz}$ **[1 mark]** $\left(\equiv 3f_0\right)$

c)
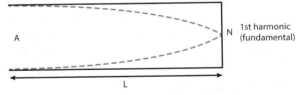
1st harmonic (fundamental)

1st harmonic: $\dfrac{\lambda}{4} = L \qquad \therefore f_0 = \dfrac{v}{\lambda} = \dfrac{v}{4L}$

$\therefore f_1 = \dfrac{330}{4 \times 1.2} = 69\,\text{Hz}$ **[1 mark]**

and $f_1 = \dfrac{330}{\left(\frac{4}{3}\right) \times 1.2} = 207\,\text{Hz}$ **[1 mark]**

$\left(\equiv 3f_2\right)$

d) Closed pipes only give odd harmonics whereas open pipes give all harmonics **[1 mark]**; open pipes usually give a more satisfying sound than closed pipes **[1 mark]**

3. a) (i) $f \propto \sqrt{T}$; if T is doubled then the frequency increases by a factor of $\sqrt{2}$ or 1.414 **[1 mark]**

(ii) $f \propto \dfrac{1}{L}$; if L is doubled then the frequency is halved or decreased by a factor of 2 **[1 mark]**

b) $L = 0.65\,\text{m}$; $T = 100\,\text{N}$; $\mu = 3.5\,\text{kg m}^{-1}$

$$f_1 = \frac{1}{2L}\sqrt{\frac{T}{\mu}} = \frac{1}{2 \times 0.65}\sqrt{\frac{100}{3.5}} = 4.1\,\text{Hz} \text{ [2 marks]}$$

c) Lighter string $\rightarrow \mu$ is smaller

$f \propto \dfrac{1}{\sqrt{\mu}}$; μ is smaller $\rightarrow \sqrt{\mu}$ is smaller **[1 mark]** and $\dfrac{1}{\sqrt{\mu}}$ is larger; therefore frequency would increase **[1 mark]**

Waves and Optics

QUICK TEST (page 62)

1. Light rays represent the direction of travel of a wavefront and are perpendicular to the wavefront
2. The normal is a line drawn at the boundary between two materials that are perpendicular or at 90° to the boundary
3. the speed of the wave and the wavelength
4. Occurs at a boundary when some of the incident wave is reflected and some is transmitted or refracted
5. $n_1 \sin\theta_1 = n_2 \sin\theta_2$
6. $n = \frac{c}{c^*}$ (ratio of speed of light in vacuum to speed of light in medium)
7. Optical density is a measure of a material's ability to pass light: an optically dense material allows light to pass, but more slowly than it would through a material that is less optically dense
8. $\sin\theta_c = \frac{n_2}{n_1}$
9. The light ray undergoes total internal reflection
10. medical endoscopes, optical fibre communications
11. a very thin and flexible core of glass or plastic that is surrounded by low refractive index cladding
12. at the core–cladding boundary
13. to avoid effects of multipath dispersion
14. Spectral dispersion occurs when white light is used; the speed of light in an optical fibre is wavelength dependent and this causes the light to become more spread out following internal reflection

PRACTICE QUESTIONS (page 63)

1. a) Refractive index is the ratio of a wave's speed in a vacuum to the wave's speed in the material **[1 mark]**

b) $n = \dfrac{\text{speed of light in air (vacuum)}}{\text{speed of light in material}}$

$\therefore 1.52 = \dfrac{3.0 \times 10^8}{c'} \Rightarrow c' = \dfrac{3.0 \times 10^8}{1.52}$

$\therefore c' = 1.97 \times 10^8 \approx 2.0 \times 10^8 \,\text{m s}^{-1}$ **[2 marks]**

c) $n_1 \sin\theta_1 = n_2 \sin\theta_2$

$n_{\text{air}} \sin\theta_1 = n_{\text{glass}} \sin\theta_2$

$\therefore \sin\theta_2 = \dfrac{n_{\text{air}}}{n_{\text{glass}}} \sin\theta_1$ **[1 mark]**

$\sin\theta_2 = \dfrac{1}{1.52} \sin 55°$

$\therefore \theta_2 = 32.6° \approx 33°$ **[1 mark]**

d) Here, $\theta_1 \equiv$ previous "θ_2" = 32.6°

$\therefore n_{\text{glass}} = \sin 32.6° = n_{\text{water}} \sin\theta_2$ **[1 mark]**

$\sin\theta_2 = \dfrac{n_{\text{glass}}}{n_{\text{water}}} \sin 32.6°$

$\therefore \sin\theta_2 = \dfrac{1.52}{1.33} \sin 32.6°$ **[1 mark]**

$\theta_2 = 38.0°$ **[1 mark]**

e) The angle of refraction is larger than the angle of incidence **[1 mark]** since the wave is travelling from an optically dense medium to a less optically dense medium **[1 mark]**

2. a) When a wave is incident from a denser medium to a less dense medium **[1 mark]** and the angle of incidence is greater than its critical angle **[1 mark]**

b) $n_1 \sin\theta_1 = n_2 \sin\theta_2$

for the critical angle, $\theta_2 = 90°$ $\therefore \sin\theta_2 = 1$

$\therefore \sin\theta_c = \dfrac{n_2}{n_1} = \dfrac{n_{\text{water}}}{n_{\text{glass}}} = \dfrac{1.33}{1.52}$ **[1 mark]**

$\therefore \theta_c = 61°$ **[1 mark]**

c) If the angle of incidence θ_i is less than the critical angle θ_c then the light ray is refracted into the water **[1 mark]** with the angle of refraction θ_r being larger than θ_i **[1 mark]**; partial internal reflection may also occur **[1 mark]**

d) $n_1 \sin\theta_1 = n_2 \sin\theta_2$

$n_{\text{glass}} \sin 45° = n_{\text{water}} \sin\theta_2$

$\therefore \sin\theta_2 = \dfrac{n_{\text{glass}}}{n_{\text{water}}} \sin 45°$ **[1 mark]**

$\therefore \theta_2 = 53.9° \approx 54°$ **[1 mark]**

3. a) A coherent bundle means that the individual fibres at each end have to be in the same relative positions in order for an accurate image to be formed **[1 mark]**

b) The cladding material has to be of a lower refractive index than the core **[1 mark]** and be equally flexible **[1 mark]**

c) $\sin\theta_c = \dfrac{n_2}{n_1} = \dfrac{1.50}{1.60}$

$\therefore \theta_c = 69.6° \approx 70°$ **[1 mark]**

d) Speed of light in glass depends on its wavelength **[1 mark]**; violet (blue) light travels more slowly than red light **[1 mark]**; difference in speed causes a pulse of white light to become broadened leading to spectral dispersion or pulse broadening **[1 mark]**; monochromatic light prevents this **[1 mark]**

Diffraction and Interference

QUICK TEST (page 66)

- $x = \frac{D\lambda}{d}$, where x is the fringe spacing, λ is the wavelength, D is the distance from the slit to the screen and d is the slit spacing
- light sources that have the same frequency (and thus wavelength) and a constant phase difference
- Fringe separation is greater for red light than for blue light
- (i) and (iii)
- Central bright fringe is twice as wide as the outer fringes that are all of the same width; the intensity of the outer fringes decreases
- 1×10^{-6} m
- $\lambda = 630$ nm
- Red laser light produces equally spaced red and dark fringes; white light produces a central white fringe with spectral bands from blue to red on either side of this central fringe

PRACTICE QUESTIONS (page 67)

a) Young's double-slit experiment demonstrates interference [1 mark] between coherent light sources [1 mark]

b) Interference fringes consisting of equally spaced bright and dark fringes [1 mark] corresponding to constructive and destructive interference [1 mark]

c) $\lambda = 445$ nm $\quad d = 0.3$ mm $= 0.3 \times 10^{-3}$ m $\quad D = 2.8$ m
$$x = \frac{D\lambda}{d} = \frac{2.8 \times 445 \times 10^{-9}}{0.3 \times 10^{-3}} \text{ [1 mark]}$$
$x = 4.15 \times 10^{-3}$ m $= 4.2$ mm [1 mark]

d) If λ increases, so does x (as both D and a remain constant) Hence, the fringe spacing increases [1 mark]

e) Only coherent sources, same frequency/wavelength [1 mark] and constant phase difference [1 mark] will produce a stable pattern of superposition

a) Spacing of the grating, $d = \dfrac{1.0 \times 10^{-3}}{650} = 1.54 \times 10^{-6}$ m [1 mark]

b) $\lambda = 635$ nm $= 635 \times 10^{-9}$ m
$$\sin\theta_1 = \frac{1 \times 635 \times 10^{-9}}{1.54 \times 10^{-6}} = 0.4123$$
$\therefore \theta_1 = 24.4°$ [1 mark]
and $\sin\theta_2 = 2 \times 0.4123 = 0.8246$
$\therefore \theta_2 = 55.5°$ [1 mark]

c) $\sin\theta_3 = 3 \times 0.4123 = 1.2369 \rightarrow$ not possible
$\theta_3 > 90° \rightarrow$ not visible [1 mark]

d) 'White' light from a filament lamp gives a continuous spectrum of colour from deep violet (350 nm) to deep red (650 nm); the diffraction pattern would therefore consist of 'bright bands' from 'violet through to red' for each order of diffraction [2 marks]

3. a) $W = \frac{2D\lambda}{a}$
W is the width of the central fringe
λ is the wavelength of light used
D is the distance between the slit and the screen
a is the width of the single slit [4 marks]

b)

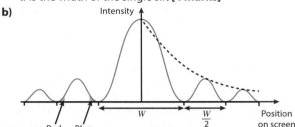

[3 marks: 1 mark for correct pattern; 1 mark for correct intensity; 1 mark for correct widths]

c) Only blue light is transmitted through the slit [1 mark] so monochromatic light is produced; a narrower diffraction pattern is obtained [1 mark]; $\lambda_{\text{blue}} \sim 445$ nm
Only red light is transmitted [1 mark]; $\lambda_{\text{red}} \sim 635$ nm; diffraction pattern is broader [1 mark]

d) $24\text{ mm} = \frac{W}{2} + \frac{W}{2} + W + \frac{W}{2} + \frac{W}{2} = 3W$
$\therefore W = \frac{24}{3} = 8\text{ mm}$
$\therefore$ width of central fringe $= 8$ mm [2 marks]

Day 5

Scalars, Vectors and Moments
QUICK TEST (page 71)

1. 10 N; 037°

2. 12 N horizontal; 21 N vertical

3.

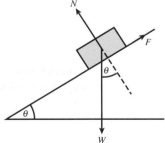

a tip-to-toe scale diagram, which will be a closed triangle

4. 1.7 m from the centre; $S = 550$ N

5. 12 N

6. The moment of the force would need to be significantly greater to destabilise the object

7. Torque is the moment (turning effect) of a couple

PRACTICE QUESTIONS (page 71)

1. a)

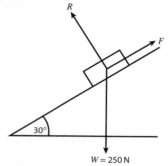

$W = 250\,\text{N}$

[3 marks: 1 mark for R; 1 mark for W; 1 mark for F]

b) perpendicular: $W\cos 30° = 250\cos 30° = 217\,\text{N}$ **[1 mark]**
parallel: $W\sin 30° = 250\sin 30° = 125\,\text{N}$ **[1 mark]**

c)

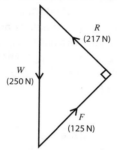

R (217 N)

W (250 N)

F (125 N)

[2 marks: 1 mark for correct triangle drawn;
1 mark for correct labelling]

2. a)

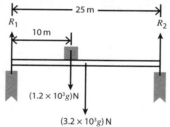

$(1.2 \times 10^3 g)\,\text{N}$

$(3.2 \times 10^3 g)\,\text{N}$

[3 marks: 1 mark for reaction force; 1 mark for weight
at centre; 1 mark for correct position of car weight]
car weight, $W_c = \left(1.2 \times 10^3\, g\right)\text{N}$
bridge weight, $W_b = \left(3.2 \times 10^3\, g\right)\text{N}$

b) car in equilibrium:
$R_1 + R_2 = W_c + W_b$
$= \left(1.2 \times 10^3\, g\right) + \left(3.2 \times 10^3\, g\right) = 4.4 \times 10^3\, g$ **[1 mark]**
$= 43.1\,\text{kN}$ **[1 mark]**
taking moments about R_1:
$\left(R_2 \times 25\right) = \left(1.2 \times 10^3\, g \times 10\right) + \left(3.2 \times 10^3\, g \times 12.5\right)$
[1 mark]
$25 R_2 = 509 \times 10^3$ [accept 510×10^3]
$R_2 = 20.4 \times 10^3 = 20.4\,\text{kN}$ **[1 mark]**
$\therefore R_1 = \left(43.1 - 20.4\right)\text{kN}$
$R_1 = 22.7\,\text{kN}$ **[1 mark]**

c) As the car moves toward the centre, the reaction forces
R_1 and R_2 would become equal **[1 mark]**

3. a)

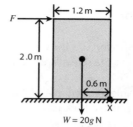

$W = 20g\,\text{N}$

[2 marks: 1 mark for correct force position; 1 mark
for correct weight position]

b) $m = 20\,\text{kg}$ $W = 20g\,\text{N}$
taking moments about X:
$F \times 2.0 = 20g \times 0.6$ **[1 mark]**
$2F = 12g$
$F \geq 6g$ **[1 mark]**
$F \geq 58.8\,\text{N} \approx 59\,\text{N}$ **[1 mark]**

c) $\tan\theta = \dfrac{0.6}{1}$ **[1 mark]**
$\therefore \theta = \tan^{-1}(0.6)$
$\theta = 31°$ **[1 mark]**

d) If the centre of gravity is lower, $\tan\theta = \dfrac{0.6}{0.5} = 1.2$
$\therefore \theta = 50°$ **[2 marks]**

e) A greater force would be required to tilt the wardrobe
beyond the new critical angle **[1 mark]**

Motion in a Straight line
QUICK TEST (page 74)

1. $6\,\text{m s}^{-1}$
2. $1\,\text{m s}^{-2}$
3. $44\,\text{m}$
4. $29\,\text{m s}^{-1}$
5. $v^2 = u^2 + 2as$
6. $21\,\text{m s}^{-1}$ in initial direction; $420\,\text{m}$
7. $0.71\,\text{s}$
8. $1.8\,\text{s}$
9. The equations of motion are independent of mass.
10. It is decelerating in one direction or accelerating in the
opposite direction.
11. by a horizontal straight line
12.

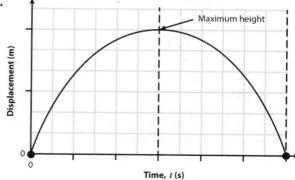

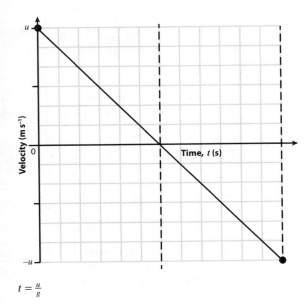

$t = \frac{u}{g}$

13. 0.1 m or 10 cm

PRACTICE QUESTIONS (page 75)

1. a)

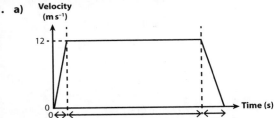

[3 marks: 1 mark for axes labelled; 1 mark for shape;
1 mark for numerical values]

b) initial acceleration: $a = \frac{\Delta v}{\Delta t} = \frac{12 - 0}{9}$ **[1 mark]** $= 1.3\,\text{m}\,\text{s}^{-2}$

[1 mark]

c) distance (i.e. area under curve) $= \frac{1}{2}(9)(12)$ **[1 mark]**

$= 54\,\text{m}$ **[1 mark]**

d) total area $= \frac{1}{2}(12)(90 + 114)$ **[1 mark]** $= 1224\,\text{m}$ **[1 mark]**

$\approx 1.22\,\text{km}$

2. a) $m = 0.1\,\text{kg}$ $T = 2.3\,\text{N}$

$T - mg = ma$ **[1 mark]**

$a = \frac{T - mg}{m}$

$a = \frac{2.3 - (0.1)(9.8)}{0.1}$ **[1 mark]**

$a = 13.2\,\text{m}\,\text{s}^{-2}$ **[1 mark]**

b) $t = 3.5\,\text{s}$

(i) $v = u + at = 0 + (13.2)(3.5)$

$v = 46.2\,\text{m}\,\text{s}^{-1} \approx 46\,\text{m}\,\text{s}^{-1}$ **[1 mark]**

(ii) $s = ut + \frac{1}{2}at^2$

$s = 0 + \frac{1}{2}(13.2)(3.5)^2$ **[1 mark]**

$= 80.9\,\text{m} \approx 81\,\text{m}$ **[1 mark]**

c) at the top of the trajectory, the speed (velocity) is 0

(i) using $u = 46.2\,\text{m}\,\text{s}^{-1}$ and $v = 0$

$v^2 = u^2 + 2as$

$0 = (46.2)^2 - 2(9.8)s$

$s = 108.9\,\text{m}$ **[1 mark]**

hence total height above ground

$= 108.9 + 80.9 = 189.8 \approx 190\,\text{m}$ **[1 mark]**

(ii) time to fall to ground:

$s = ut + \frac{1}{2}at^2$

$189.75 = 0 + \frac{1}{2}(9.8)t^2$ **[1 mark]**

$t = 6.2\,\text{s}$ **[1 mark]**

3. a) total thrust $= 4 \times 220 = 880\,\text{kN}$ **[1 mark]**

b)

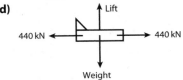

resultant force $= 880 - 50 = 830\,\text{kN}$ **[1 mark]**

$R = ma \quad \therefore a = \frac{R}{m} = \frac{830 \times 10^3}{320 \times 10^3}$ **[1 mark]**

$a = 2.59 \approx 2.6\,\text{m}\,\text{s}^{-2}$ **[1 mark]**

c) $t = 15\,\text{min} = 15 \times 60 = 900\,\text{s}$ $v = 250\,\text{m}\,\text{s}^{-1}$ $u = 0\,\text{m}\,\text{s}^{-1}$

$v = u + at \rightarrow a = \frac{v - u}{t}$ **[1 mark]**

$a = \frac{250 - 0}{900} = 0.28\,\text{m}\,\text{s}^{-2}$ **[1 mark]**

d)

440 kN ← | → 440 kN   ↑ Lift   ↓ Weight

drag $= 440\,\text{kN}$ **[1 mark]**

lift $=$ weight $= mg$

$= 300 \times 10^3 \times 9.8$ **[1 mark]**

$= 2.94 \times 10^6\,\text{N}$ (or 2.94 MN) **[1 mark]**

e) $a = -2.0\,\text{m}\,\text{s}^{-2}$ $u = 45\,\text{m}\,\text{s}^{-1}$ $v = 0\,\text{m}\,\text{s}^{-1}$

(i) $v = u + at \rightarrow t = \frac{v - u}{a} = \frac{0 - 45}{-2}$ **[1 mark]** $= 22.5\,\text{s}$

[1 mark]

(ii) $v^2 = u^2 + 2as \rightarrow s = \frac{v^2 - u^2}{2a} = \frac{0 - 45^2}{-4}$ **[1 mark]**

$= 506\,\text{m}$ **[1 mark]**

Projectile Motion
QUICK TEST (page 78)

1. $0\,\text{m}\,\text{s}^{-1}$ or zero

2. $8\,\text{m}\,\text{s}^{-1}$

3. yes

4. yes, because the acceleration due to gravity is constant and the horizontal and vertical motions are independent of each other

5. force due to gravity, also possibly air resistance or drag

6. the acceleration that acts vertically downwards due to gravity

7. $-12\,\text{ms}^{-1}$

8. the greatest height, above its projection point, attained during its trajectory

9. the total time the projectile is in the air

10. $45°$

11. (i) $2.5\,\text{s}$ (ii) $50\,\text{m}$ (iii) $31\,\text{m s}^{-1}$

12. (i) $2.4\,\text{s}$ (ii) $50\,\text{m}$ (iii) $8.8\,\text{m}$

PRACTICE QUESTIONS (page 79)

1. a) $v^2 = u^2 + 2as$

$$s = \frac{v^2 - u^2}{-2g} = \frac{0 - 25}{-19.6}\ \text{[1 mark]}$$

$s = 1.28\,\text{m}$ **[1 mark]**

∴ height above ground $= 2 + 1.28 = 3.38\,\text{m}$

$\approx 3.4\,\text{m}$ **[1 mark]**

b) $v = u + at \rightarrow v = u - gt$ **[1 mark]**

$$\therefore t = \frac{u}{g} = \frac{5}{9.8} = 0.51\text{s}\ \text{[1 mark]}$$

c) If air resistance cannot be ignored, the resistance would slow the ball down and the maximum height would be lower **[1 mark]**

d) $u = 0\,\text{ms}^{-1}$

$v = ?$

$a = g = 9.8\,\text{ms}^{-2}$

$s = 3.38\,\text{m}$

$v^2 = u^2 + 2gs$

$= 0^2 + 2(9.8)(3.38)$ **[1 mark]**

$v = \sqrt{66.25} = 8.139 \approx 8.14\,\text{ms}^{-1} \approx 8.1\,\text{m s}^{-1}$ **[1 mark]**

2.

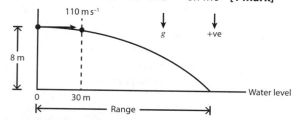

a) vertically: $u = 0$

$v = ?$

$s = 8\,\text{m}$

$a = g = 9.8\,\text{ms}^{-2}$

$t = ?$

$$s = ut + \frac{1}{2}gt^2$$

$8 = 0 + \frac{1}{2}(9.8)t^2$ **[1 mark]**

$$t = \sqrt{\frac{8}{4.9}} = 1.28\text{s} \approx 1.3\text{s}\ \text{[1 mark]}$$

b) range $= s = ut = 110 \times 1.28 = 140.8\,\text{m}$

$\approx 141\,\text{m}$ **[1 mark]**

c) time to reach 30 m away: $t = \dfrac{30}{110} = 0.273\text{s}$ **[1 mark]**

in this time, the cannonball drops a distance s:

$$s = \frac{1}{2}gt^2 = 0.36\,\text{m}\ \text{[1 mark]}$$

∴ height above water is $8 - 0.36 = 7.64\,\text{m}$

$\approx 7.6\,\text{m}$ **[1 mark]**

3.

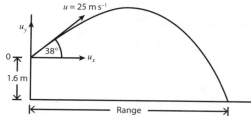

a) Ignore air resistance and spin **[1 mark]**

b) $u_x = 25\cos 38° = 19.70 \approx 19.7\,\text{m s}^{-1}$ **[1 mark]**

$u_y = 25\sin 38° = 15.39 \approx 15.4\,\text{m s}^{-1}$ **[1 mark]**

c) maximum height is reached when $v_y = 0$

vertically: $u_y = 15.4\,\text{m s}^{-1}$

$v_y = 0$

$s_y = ?$

$a = -g = -9.8\,\text{m s}^{-2}$

$t = ?$

$v^2 = u^2 - 2gs$

$$s = \frac{u^2 - v^2}{2g} \rightarrow s_y = \frac{u_y^2 - v_y^2}{2g}\ \text{[1 mark]}$$

$$s_y = \frac{15.4^2}{19.6} = 12.1\,\text{m}\ \text{[1 mark]}$$

∴ hammer is $12.1 + 1.6\,\text{m}$ above ground $= 13.7\,\text{m}$ **[1 mark]**

time of flight:

vertically: $u_y = 15.4\,\text{m s}^{-1}$

$s_y = -1.6\,\text{m}$

$a = -g = -9.8\,\text{m s}^{-2}$

use $s_y = u_y t - \dfrac{1}{2}gt^2$ **[1 mark]**

$-1.6 = 15.4t - 4.9t^2$ **[1 mark]**

$4.9t^2 - 15.4t - 1.6 = 0$

solve quadratic equation using formula

$$\frac{-b \pm \sqrt{b^2 - 4ac}}{2a} = \frac{15.4 \pm \sqrt{(15.4)^2 - (-4 \times 4.9 \times 1.6)}}{9.8} = 3.2$$

$t = 3.2\text{ s}$ **[1 mark]** (ignore $t = -0.10\text{s}$)

d) range $s = u_x t = 19.7 \times 3.2$

$= 63.04\,\text{m} \approx 63\,\text{m}$ **[1 mark]**

f) $\theta > 38°$ but $\leq 45°$ to give a greater range **[1 mark]**

Energy, Work and Power

QUICK TEST (page 82)

1. $9000\,\text{J}$

2. $310\,000\,\text{J}$ or $3.1 \times 10^5\,\text{J}$ or $310\,\text{kJ}$

3. (i) $5.0\,\text{J}$ (ii) $5.0\,\text{J}$ (iii) $1.3\,\text{m}$

4. $2100\,\text{J}$ or $2.1\,\text{kJ}$

5. $4.8 \times 10^6\,\text{W}$

6. (i) 12% (ii) energy is lost through heating

7. 166 MW

8. (i) energy; loss of elastic potential energy = gain in kinetic energy (ii) $11.2\,\mathrm{m\,s^{-1}}$

PRACTICE QUESTIONS (page 83)

1. **a)** $h = 65\,\mathrm{m}$ $\quad m = 1.0\,\mathrm{kg}$

$\Delta E_\mathrm{p} = mgh = 1.0 \times 9.8 \times 65 = 637\,\mathrm{J}$ **[1 mark]**

$\approx 640\,\mathrm{J}$ **[1 mark]**

b) flow rate $= 4.4 \times 10^7\,\mathrm{kg\,h^{-1}}$

$\Delta E_\mathrm{p}\,(\text{total}) = 637 \times 4.4 \times 10^7\,\mathrm{J\,h^{-1}}$ **[1 mark]**

$= \dfrac{637 \times 4.4 \times 10^7}{3600}\,\mathrm{W}$ **[1 mark]**

$= 7.786 \times 10^6 \approx 7.8\,\mathrm{MW}$ **[1 mark]**

c) $\eta = \dfrac{5.5}{7.8} \times 100$ **[1 mark]** $= 70.5\% \approx 71\%$ efficient **[1 mark]**

2. $m = 70\,\mathrm{kg}$ $\quad h = 41.4\,\mathrm{km} = 41.4 \times 10^3\,\mathrm{m}$

loss of height in free fall $= 37.6\,\mathrm{km} = 37.6 \times 10^3\,\mathrm{m}$

a) $\Delta E_\mathrm{p}\,(\text{loss}) = mg\Delta h = 70 \times 9.8 \times 37.6 \times 10^3$ **[1 mark]**

$= 25.79 \times 10^6 \approx 25.8 \times 10^6\,\mathrm{J}$

$= 25.8\,\mathrm{MJ}$ **[1 mark]**

b) $\Delta E_\mathrm{p}\,(\text{loss}) = \Delta E_\mathrm{k}\,(\text{gain}) = \dfrac{1}{2}mv^2$

$v = \sqrt{\dfrac{2\Delta E_\mathrm{p}}{m}}$ **[1 mark]** $= \sqrt{\dfrac{2 \times 25.8 \times 10^6}{70}}$

$v = 858.6 \approx 860\,\mathrm{m\,s^{-1}}$ **[1 mark]**

$\therefore\, v = \dfrac{860 \times 3600}{1000}\,\mathrm{km\,h^{-1}}$

$= 3090 \approx 3100\,\mathrm{km\,h^{-1}}$ **[1 mark]**

c) There would have been some air resistance, which would be significant at lower altitudes **[1 mark]**

Air resistance $\propto$ velocity^2; air resistance $\propto$ surface area **[1 mark]**

d) $1300\,\mathrm{km\,h^{-1}} \approx 360\,\mathrm{m\,s^{-1}}$

$\Delta W = \Delta E = \dfrac{1}{2}m\Delta(v^2) = \dfrac{1}{2} \times 70 \times \left(860^2 - 360^2\right)$ **[1 mark]**

$\Delta W = 21.35 \times 10^6\,\mathrm{J} = 21.4\,\mathrm{MJ}$ **[1 mark]**

e) $\Delta W = F \times s$ $\quad \therefore\, F = \dfrac{\Delta W}{s} = \dfrac{\Delta W}{\Delta h}$ **[1 mark]**

$F_\mathrm{mean} = \dfrac{21.4 \times 10^6}{37.6 \times 10^3} = 569.1\,\mathrm{N} \approx 570\,\mathrm{N}$ **[1 mark]**

3. $m = 120\,000\,\mathrm{kg}$ $\quad v = 220\,\mathrm{m\,s^{-1}}$

a) weight $= mg = 120\,000 \times 9.8 = 1.176 \times 10^6\,\mathrm{N}$ **[1 mark]**

up-thrust $\dot= 1.176 \times 10^6\,\mathrm{N}$

$\approx 1.2 \times 10^6\,\mathrm{N}$ **[1 mark]**

b) constant speed thus no acceleration

$\therefore$ thrust $=$ drag

$= 80\,\mathrm{kN}$ **[1 mark]**

power $= Fv$

$= 80 \times 10^3 \times 220$

$= 17.6 \times 10^6\,\mathrm{W}$ **[1 mark]**

$\therefore$ power in each engine $= \dfrac{1}{4} \times 17.6 \times 10^6$

$= 4.4 \times 10^6\,\mathrm{W}$ **[1 mark]**

c) Reducing the power would mean that the drag then exceeded the thrust **[1 mark]**

From Newton's Second Law ($F = ma$), this would result in the aeroplane decelerating **[1 mark]**

Momentum and Collisions
QUICK TEST (page 86)

1. $0.5\,\mathrm{kg\,m\,s^{-1}}$

2. $0.15\,\mathrm{kg\,m\,s^{-1}}$

3. Force is equal to the rate of change of momentum, $F = \dfrac{\Delta p}{\Delta t}$

4. $-9.3\,\mathrm{N\,s}$ or $-9.3\,\mathrm{kg\,m\,s^{-1}}$

5. (i) $F = \dfrac{\Delta p}{\Delta t}$ (ii) 775 N

6. (i) $-3.2 \times 10^{-23}\,\mathrm{N\,s}$ (ii) $2.1 \times 10^{-13}\,\mathrm{N}$

7. $3.75\,\mathrm{m\,s^{-1}}$

8. Crumple zones, which extend the contact time during an impact and hence reduce the impact force on the passengers

9. $-0.07\,\mathrm{m\,s^{-1}}$ (opposite direction to shell)

10. change in momentum; N s or kg m s^{-1}

PRACTICE QUESTIONS (page 87)

1. **a)** $m_\mathrm{e} = 6.5 \times 10^4\,\mathrm{kg}$ $\quad u_\mathrm{e} = 0.4\,\mathrm{m\,s^{-1}}$ $\quad m_\mathrm{w} = 8.5 \times 10^4\,\mathrm{kg}$

$p_\mathrm{e} = m_\mathrm{e}u_\mathrm{e} = 6.5 \times 10^4 \times 0.4$ **[1 mark]**

$= 2.6 \times 10^4\,\mathrm{kg\,m\,s^{-1}}$ **[1 mark]**

b) $m_\mathrm{e}u_\mathrm{e} + 0 = \left(m_\mathrm{e} + m_\mathrm{w}\right)V$ **[1 mark]**

$2.6 \times 10^4 = \left(15 \times 10^4\right)V$

$\therefore\, V = 0.17\,\mathrm{m\,s^{-1}}$ **[1 mark]** in the same direction as the engine **[1 mark]**

2. **a)** kinetic energy **[1 mark]** and momentum **[1 mark]**

b) (i) $p_\mathrm{w} = m_\mathrm{w}u_\mathrm{w} = 0.1 \times 5 = 0.5\,\mathrm{kg\,m\,s^{-1}}$ **[1 mark]**

(ii) $m_\mathrm{w}u_\mathrm{w} + m_\mathrm{r}u_\mathrm{r} = m_\mathrm{w}v_\mathrm{w} + m_\mathrm{r}v_\mathrm{r}$

$0.5 + 0 = \left(0.1 + 1.5\right) + \left(0.1 \times v_\mathrm{r}\right)$ **[1 mark]**

$0.5 = 0.15 + 0.1v_\mathrm{r}$

$v_\mathrm{r} = \dfrac{0.5 - 0.15}{0.1}$ **[1 mark]**

$= 3.5\,\mathrm{m\,s^{-1}}$ **[1 mark]** in the same direction as the white ball

c) $F = \dfrac{\Delta p}{\Delta t} = \dfrac{-2(0.1)(3.5)}{0.07}$ **[1 mark]**

$F = -10\,\mathrm{N}$ **[1 mark]**

3. **a)** $F_\mathrm{mean} \approx 20 \times 10^3\,\mathrm{N}$ **[1 mark]**

b) impulse $I = \Delta p = F\Delta t$

$= 20 \times 10^3 \times 7 \times 10^{-4}$ **[1 mark]**

$= 14\,\mathrm{N\,s}$ **[1 mark]**

c) change in momentum $\Delta p = mv - mu$

$\Delta p = 0.16(v - u)$ **[1 mark]**

$14 = 0.16(v - u)$

$v - u = 87.5$ **[1 mark]**

$\therefore\, v = 87.5 + u$

but $u = -40\,\mathrm{m\,s^{-1}}$

$\therefore\, v = 47.5\,\mathrm{m\,s^{-1}} \approx 48\,\mathrm{m\,s^{-1}}$ **[1 mark]** in opposite direction to initial velocity **[1 mark]**

4. a) In an inelastic collision there is a loss of kinetic energy **[1 mark]**

b) $m_1u_1 + m_2u_2 = m_1v_1 + m_2v_2$

$(1.5 \times 6.5) + (0.5 \times [-4.0]) = (1.5v_1) + (0.5 \times 3.5)$ **[1 mark]**

$9.75 - 2.0 = 1.5v_1 + 1.75$

$1.5v_1 = 6.0$

$v_1 = 4.0\,\text{ms}^{-1}$ **[1 mark]** in the same direction as before the collision **[1 mark]**

c) $\frac{1}{2}m_1u_1^2 + \frac{1}{2}m_2u_2^2 = \frac{1}{2}m_1v_1^2 + \frac{1}{2}m_2v_2^2 + \Delta E$ **[1 mark]**

$\frac{1}{2}(1.5)(6.5^2) + \frac{1}{2}(0.5)(-4)^2 = \frac{1}{2}(1.5)(4^2) +$
$$\frac{1}{2}(0.5)(3.5^2) + \Delta E$$

$3.17 + 4 = 12 + 3.1 + \Delta E$ **[1 mark]**

$35.7 = 15.1 + \Delta E$

$\therefore \Delta E = 20.6\,\text{J} \approx 21\,\text{J}$ **[1 mark]**

21 J of energy is transferred into other forms of energy **[1 mark]**

Day 6

Bulk Properties of Materials
QUICK TEST (page 90)

1. when a material returns to its original length when the load has been removed
2. The extension is directly proportional to the load (force) applied
3. It is the straight-line part of the graph
4. a material that shows a large plastic region
5. the stiffness constant or spring constant
6. a spring
7. steeper gradient
8. dislocations
9. $\text{kg}\,\text{m}^{-3}$
10. force that is pulling or stretching a material
11. the point beyond which Hooke's Law is no longer obeyed
12. the point beyond which plastic deformation occurs
13. $k = \frac{F}{\Delta l} = \frac{72}{10} = 720\,\text{Nm}^{-1}$
14. $k_{\text{eff}} = 2 \times 720 = 1440\,\text{Nm}^{-1}$

PRACTICE QUESTIONS (page 91)

1. a) The density of a substance is its mass per unit volume or $\rho = \frac{m}{V}$ **[1 mark]**

b) $V_{\text{copper}} = 0.7 \times 2.2 \times 10^{-5} = 1.54 \times 10^{-5}\,\text{m}^3$ and

$V_{\text{zinc}} = 0.3 \times 2.2 \times 10^{-5} = 6.66 \times 10^{-6}\,\text{m}^3$ **[1 mark]**

$m_{\text{copper}} = \rho_{\text{copper}} \times V_{\text{copper}} = 8.9 \times 10^3 \times 1.54 \times 10^{-5}$
$$= 0.137\,\text{kg}\ \textbf{[1 mark]}$$

$m_{\text{zinc}} = 7.1 \times 10^3 \times 6.66 \times 10^{-6} = 0.047\,\text{kg}$ **[1 mark]**

c) $m_{\text{brass}} = 0.137 + 0.047 = 0.184\,\text{kg}$

$\therefore \rho_{\text{brass}} = \frac{m_{\text{brass}}}{V_{\text{brass}}} = \frac{0.184}{2.2 \times 10^{-5}} = 8362\,\text{kgm}^{-3}$

$= 8400\,\text{kgm}^{-3}$ (2 s.f.) **[2 marks]**

2. a) The force **[1 mark]** needed to stretch a spring/wire is directly proportional to the extension **[1 mark]** of the spring/wire

b) X is the limit of proportionality, Y is the elastic limit and Z is the point of fracture **[2 marks for all three correct; 1 mark for any two correct]**

c) Gradient of OX represents the stiffness constant or spring constant **[1 mark]**

d) Material A shows ductility **[1 mark]**; the plastic region **[1 mark]**

e) brittle **[1 mark]**; greater stiffness constant **[1 mark]**; small plastic region **[1 mark]**; ceramic or glass **[1 mark]**

3. a) $F = 8\,\text{N}$ $\Delta l = 5\,\text{cm} = 0.05\,\text{m}$

$F = k\Delta l \therefore k = \frac{F}{\Delta l} = \frac{8}{0.05}$ **[1 mark]** $= 160\,\text{Nm}^{-1}$ **[1 mark]**

b) $\Delta l = 25\,\text{cm} = 0.25\,\text{m}$

$F = k\Delta l = 160 \times 0.25 = 40\,\text{N}$ **[1 mark]**

c) three springs in parallel gives $k_{\text{eff}} = k + k + k = 3k$

$= 480\,\text{Nm}^{-1}$ **[1 mark]**

$\therefore \Delta l = \frac{F}{k_{\text{eff}}}$ **[1 mark]** $= \frac{100}{480} = 0.208\,\text{m} \approx 21\,\text{cm}$ **[1 mark]**

d) that the elastic limit is not reached/exceeded **[1 mark]**

Stress, Strain and the Young Modulus
QUICK TEST (page 94)

1. Tensile stress is the force per unit area of a material being stretched whereas compressive stress is the force per unit area when the material is squashed
2. ratio of extension to original length
3. Nm^{-2} or pascals (Pa)
4. $E = \dfrac{\text{stress}}{\text{strain}}$
5. the maximum stress value withstood by a material before breaking
6. necking
7. by its ultimate tensile stress value
8. by a steeper line within the elastic region
9. the continuous deformation of a material under constant stress below its yield point
10. Ceramic, owing to its higher value of the Young modulus or because it takes a greater stress to cause fracture
11. $2.1 \times 10^{11}\,\text{Pa}$ or 210 GPa
12. extensive (almost linear) elastic region and almost negligible plastic region

PRACTICE QUESTIONS (page 95)

1. a) (i) stiffer: material A **[1 mark]**; steeper initial gradient **[1 mark]**

(ii) tougher: material B **[1 mark]**; can withstand higher stress and higher strain before fracturing **[1 mark]**

(iii) stronger: material B **[1 mark]**; greater UTS **[1 mark]**

(iv) more ductile: material B **[1 mark]**; larger plastic region **[1 mark]**

b)

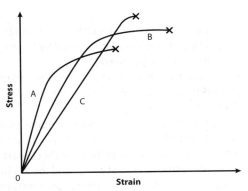

Material C has lower gradient **[1 mark]**, greater UTS **[1 mark]** and fractures soon after the elastic region **[1 mark]**

2. **a)** **(i)** plastic flow: B **[1 mark]**
   (ii) necking: C **[1 mark]**
   (iii) elastic deformation: A **[1 mark]**

   b) X: limit of proportionality (or elastic limit) **[1 mark]**
   Y: ultimate tensile stress (UTS) **[1 mark]**

   c) $E = \dfrac{\text{stress}}{\text{strain}} = \dfrac{200\,\text{MPa}}{0.002}$ **[1 mark]**

   $E = \dfrac{200 \times 10^6}{0.002}$ **[1 mark]** $= 1.0 \times 10^{11}\,\text{Pa}$

   $E = 100\,\text{GPa} = 100\,000\,\text{MPa}$ **[1 mark]**

3. $d = 3.0\,\text{mm} = 3.0 \times 10^{-3}\,\text{m}$

   a) $A = \dfrac{\pi}{4}d^2 = \dfrac{\pi}{4}\left(3.0 \times 10^{-3}\right)^2\,\text{m}^2$ **[1 mark]**

   $A = 7.07 \times 10^{-6}\,\text{m}^2 \approx 7.1 \times 10^{-6}\,\text{m}^2$ **[1 mark]**

   b) $F = 1750\,\text{N}$

   tensile stress $= \dfrac{F}{A} = \dfrac{1750}{7.07 \times 10^{-6}}$ **[1 mark]**

   $= 2.48 \times 10^8\,\text{Pa} \approx 2.5 \times 10^8\,\text{Pa}$ **[1 mark]**

   c) $E = 210\,\text{GPa} = 210 \times 10^9\,\text{Pa}$ $\qquad l = 45\,\text{m}$

   $E = \dfrac{(F/A)}{(\Delta l / l)} \rightarrow \Delta l = \dfrac{(F/A)l}{E}$ **[1 mark]**

   $\Delta l = \dfrac{2.48 \times 10^8 \times 45}{210 \times 10^9}$ **[1 mark]** $= 0.053\,\text{m}$

   $\Delta l = 53\,\text{mm}$ **[1 mark]**

Strain Energy and Toughness
QUICK TEST (page 98)

1. the average force (load) multiplied by the extension produced, i.e. $\frac{1}{2}F\Delta l$
2. strain energy
3. because it only appears when the load is removed
4. from the area under the curve
5. the strain energy density
6. $5.6 \times 10^{-3}\,\text{J}$
7. 35 mm
8. $6.0 \times 10^4\,\text{J m}^{-3}$ or $60\,\text{kJ m}^{-3}$
9. the amount of internal energy retained when unloaded
10. the strain energy required to produce the permanent deformation

PRACTICE QUESTIONS (page 99)

1. **a)** $E = \dfrac{\text{stress}}{\text{strain}}$ in elastic region $= \dfrac{2.8 \times 10^8}{0.5}$ **[1 mark]**

   $= 0.56 \times 10^9\,\text{Pa}$ **[1 mark]**

   $E = 0.56\,\text{GPa}$ or $E = 560\,\text{MPa}$ **[1 mark]**

   b) UTS $= 3.8 \times 10^8\,\text{Pa}$ **[1 mark]**

   c) energy stored per unit volume = area under curve
   each 1 cm square $= \left(0.5 \times 10^8\right) \times \left(0.25 \times 10^{-3}\right) = 12500\,\text{J m}^{-3}$ **[1 mark]**
   Number of 1 cm squares = 48 **[1 mark]**
   $\therefore E = 12500 \times 48\,\text{J}$
   $= 600 \times 10^3\,\text{J}$
   $= 600\,\text{kJ m}^{-3}$ **[1 mark]**

   d) tough **[1 mark]**; has a high fracture energy **[1 mark]**

2. **a)**

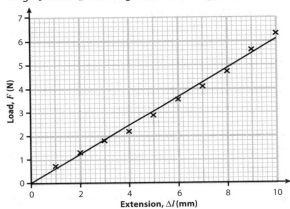

[4 marks: 1 mark for axes correctly labelled, 1 mark for all points plotted correctly, 1 mark for line of best fit drawn, 1 mark for gradient calculated using large triangle]

   $k = \dfrac{6.1}{10 \times 10^{-3}}$ **[1 mark]** $= 610\,\text{Nm}^{-1}$ **[1 mark]**

   b) work done = area under curve = area of triangle **[1 mark]**
   $= \dfrac{1}{2}\text{base} \times \text{height} = \dfrac{1}{2}\Delta l \times F = \dfrac{1}{2}F\Delta l$ **[1 mark]**

   c) area under graph $= \dfrac{1}{2} \times \left(10 \times 10^{-3}\right) \times 6.1$ **[1 mark]**
   $= 0.031\,\text{J}$ **[1 mark]**

3. **a)** $E = \dfrac{\text{stress}}{\text{strain}}$ in initial region $= \dfrac{0.2 \times 10^9}{0.03}$ **[1 mark]**
   $E = 6.7 \times 10^9\,\text{Pa}$ **[1 mark]** $= 6.7\,\text{GPa}$

   b) strain energy density $= \dfrac{1}{2}(\text{stress})(\text{strain})$

   $= \dfrac{1}{2}\left(0.2 \times 10^9\right)(0.03)$ **[1 mark]**

   $= 3.0 \times 10^6 = 3.0\,\text{MJ m}^{-3}$ **[1 mark]**

   c) extensive plastic behaviour **[1 mark]**; almost linear stress–strain curve up to fracture point **[1 mark]**

d) Approximating the graph to a straight line, taking point (0, 0) and (0.21, 1.4 GPa) **[1 mark]** gives

$$\text{strain energy density} = \frac{1}{2}(\text{stress})(\text{strain})$$

$$= \frac{1}{2}(0.21)(1.4 \times 10^9) \text{ [1 mark]}$$

$$= 147 \times 10^6 = 147 \, \text{MJ m}^{-3} \text{ [1 mark]}$$

e) volume = cross-sectional area × length

$$= \frac{\pi}{4}(4.0 \times 10^{-6})^2 \times (50 \times 10^{-3}) \text{ [1 mark]}$$

$$= 6.3 \times 10^{-13} \, \text{m}^3 \text{ [1 mark]}$$

$$\therefore \text{strain energy stored} = 147 \times 10^6 \, \text{MJ m}^{-3} \times 6.3 \times 10^{-13} \, \text{m}^3$$

[1 mark]

$$= 9.2 \times 10^{-5} \, \text{J} \text{ [1 mark]}$$

Microstructure of Materials
QUICK TEST (page 102)

1. because they have a strong ionic crystal structure and so cracks do not propagate through them
2. a structure that has no long-range order
3. long-chain molecules
4. the cross-linking of chains
5. because they have strong ionic and covalent bonds between atoms forming a very rigid structure
6. The movement of dislocations produces slip
7. the formation of new dislocations that interact with each other
8. because cracks and other flaws limit their strength
9. because of the presence and movement of dislocations within the material
10. because they have no long-range order; they have an amorphous structure like a liquid
11. [any three from] grain boundaries, interstitial and substitutional impurity atoms, self-interstitial atoms, vacancies, introducing more dislocations through work hardening

PRACTICE QUESTIONS (page 103)

1. **a)** material A: stiff, brittle and strong, e.g. brick/ceramic **[1 mark]**
   material B: metallic, e.g. copper wire **[1 mark]**
   b) **(i)** **Elastic deformation**
   The proportional extension of a material up to the elastic limit **[1 mark]**, whereupon it returns to its original length once the strain is removed **[1 mark]**; bonds between atoms act like elastic springs
   Plastic deformation
   Beyond the elastic limit, permanent deformation occurs **[1 mark]**; the material extends rapidly for small increases in stress until it fractures **[1 mark]**
   (ii) **Ductile**
   Materials are described as ductile if the relative movement of atoms is easily achieved **[1 mark]**; results in a reduction of the cross-sectional area and eventually fracture due to necking **[1 mark]**

Brittle
Materials that show very little or no plastic region **[1 mark]**; necking is not seen and fracture occurs instantaneously **[1 mark]**

2. **a)** A dislocation is a line defect in a crystalline solid **[1 mark]** that explains the movement of atoms in ductile metals **[1 mark]**
   b) Dislocation takes the form of an incomplete plane of atoms **[1 mark]**; if the stress is maintained then the dislocation moves through the material **[1 mark]**; only one bond is broken at a time and the amount of energy required for movement is significantly reduced **[1 mark]**
   c) [Any two from]
   The introduction of 'foreign' atoms into the lattice **[1 mark]**; these disturb the lattice and in so doing hinder the progress of dislocations and their movement through the lattice **[1 mark]**
   and/or
   By heating/cooling that introduces more grains **[1 mark]** and therefore more grain boundaries; grain boundaries inhibit and stop the movement of dislocations **[1 mark]**
   and/or
   By work hardening that introduces more dislocations **[1 mark]** into a material and these become entangled with other dislocations, preventing further movement **[1 mark]**
   [4 marks: 1 mark for each correct point]

3.

Feature	Metal	Ceramic
Young modulus	Steep gradient of the stress–strain curve, indicating a high value of the Young modulus, e.g. 100 GPa	Very steep gradient of the stress–strain curve, indicating a very high value of the Young modulus, e.g. 350 GPa
Elastic deformation	Not as strong as ceramic; shows elastic region	High stress for little strain; significant elastic region
Plastic deformation	Extensive plastic region, showing yield point, UTS and eventually fracture	Very limited plastic region; shows brittle behaviour
Fracture	After UTS, cross-section reduces, resulting in necking and eventually fracture	No necking; no change in cross-section; spontaneous fracture

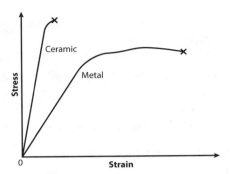

[6 marks: 1 mark for each correct row in the table; 1 mark for each correctly plotted line on the graph]

Day 7

Direct Current, Ohm's Law and I–V Characteristics

QUICK TEST (page 106)

1. a source of electrical energy
2. $\Delta Q = I\Delta t$ t or $I = \dfrac{\Delta Q}{\Delta t}$
3. 4.5 C
4. the flow of charge from positive to negative and opposite to the direction of electron flow
5. the difference in potential across the ends of a device or the energy required to move unit charge through a device or the work done per unit charge
6. the ratio of the potential difference across a device to its current; $R = \dfrac{V}{I}$
7. 300 kΩ
8. 30 W or 30 J s^{-1}
9. (i) A s (ii) C s^{-1}
10. Resistance increases with temperature owing to more collisions between charge carriers and atoms
11. The device obeys Ohm's Law and the current–voltage graph is a straight line through the origin; current is directly proportional to potential difference
12. Resistance decreases
13. Allows current to flow in one direction only

PRACTICE QUESTIONS (page 107)

1. a)

[2 marks: 1 mark for plotting correct curve; 1 mark for correct labels of light and dark]

b)

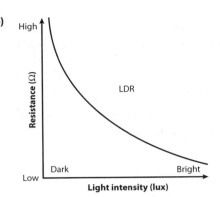

[2 marks: 1 mark for plotting correct curve; 1 mark for correct labels of dark and bright]

c) LDRs are made from semiconducting materials [1 mark]; when the light diminishes, the number of 'free' charge carriers diminishes [1 mark]; the current decreases and hence the resistance increases according to $I = \dfrac{V}{R}$ [1 mark]

2. a)

[3 marks: 1 mark for correctly labelled axes; 1 mark for points plotted correctly; 1 mark for straight line drawn]

b) Slope represents the reciprocal of the resistance of the component, $\dfrac{1}{R}$ [1 mark]

c) slope $= \dfrac{\Delta I}{\Delta V} = \dfrac{0.027}{8}$ $\therefore R = \dfrac{8}{0.027} = 296.3\,\Omega \approx 300\,\Omega$

[2 marks]

d) most probably a metal conductor at constant temperature [1 mark] that obeys Ohm's Law, i.e. an ohmic conductor [1 mark]

e) At an elevated (but constant) temperature, expect a similar straight-line response [1 mark] but with a much shallower gradient, i.e. R is larger [1 mark]

f) In metals at elevated temperatures, the atoms vibrate with greater amplitude [1 mark]; collisions between free electrons and ion cores are now more frequent [1 mark] and this gives an increase in the amount of kinetic energy transferred from the electrons; the result is an increase in the resistance [1 mark]

3. a) For a filament lamp, e.g. a standard tungsten lamp, electrical energy is converted into light and heat **[1 mark]**; a higher current leads to a higher temperature **[1 mark]**, which leads to a higher resistance **[1 mark]**; the I–V characteristics show the ratio $\left(\dfrac{I}{V}\right)$ decreasing and therefore the resistance increasing; therefore it is not an ohmic conductor **[1 mark]**

b)

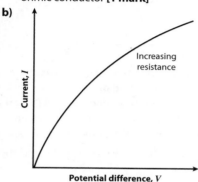

[2 marks: 1 mark for correctly plotted graph; 1 mark for correct axes labels]

c) The resistance increases with increasing temperature **[1 mark]**; as the temperature increases, so does the frequency of collisions between the electrons and the positive ion cores of the metal atoms **[1 mark]**; because these ion cores vibrate with greater amplitude with increasing temperature **[1 mark]**

d) Set up the circuit as shown; increase the current using the variable resistor and record both V and I at the same time

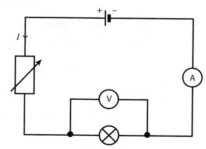

[3 marks: 1 mark for description; 2 marks for correctly drawn circuit diagram]

Resistance, Resistivity, Conductivity and Superconductivity
QUICK TEST (page 110)

1. 176 Ω

2. 13 Ω

3. At any junction in a circuit, the sum of the current flowing into a junction is the same as the sum of the current flowing away from it (conservation of charge/current)

4. It adjusts the current flow so that different values of current and voltage can be recorded

5. resistivity of its material, length and cross-sectional area

6. $R = \dfrac{\rho l}{A}$ or resistance is directly proportional to resistivity

7. $\Omega\,\mathrm{m}$

8. superconductor

9. critical temperature or transition temperature

PRACTICE QUESTIONS (page 111)

1. a) $\dfrac{1}{R_{\text{total}}} = \dfrac{1}{1000} + \dfrac{1}{550} = \dfrac{31}{11000}$

$\therefore R_{\text{total}} = \dfrac{11000}{31} = 354.8\,\Omega = 355\,\Omega$ **[2 marks]**

b) $V = IR \rightarrow I = \dfrac{V}{R_{\text{total}}} = \dfrac{12}{355} = 0.0338\,\text{A}$

$\therefore I \approx 0.034\,\text{A or } 34\,\text{mA}$ **[2 marks]**

c) $R_{\text{total}} = 355 + 500 = 855\,\Omega$

$\therefore I = \dfrac{V}{R_{\text{total}}} = \dfrac{12}{855} = 0.0140\,\text{A} \approx 14\,\text{mA}$

Additional resistance in series reduces the current drawn from the battery because the overall circuit resistance is greater

[3 marks: 2 marks for calculations; 1 mark for comment]

2. a) $R_1 = 200\,\Omega + 500\,\Omega = 700\,\Omega$

$\dfrac{1}{R_{\text{total}}} = \dfrac{1}{700} + \dfrac{1}{1000} = \dfrac{17}{7000}$

$\therefore R_{\text{total}} = \dfrac{7000}{17} = 411.8\,\Omega \approx 410\,\Omega$ **[2 marks]**

b) $V = IR_{\text{total}} \rightarrow I = \dfrac{V}{R_{\text{total}}} = \dfrac{9}{412}$

$\therefore I = 0.0218\,\text{A} \approx 22\,\text{mA}$ **[2 marks]**

c) $P = IV = 0.0218 \times 9 = 0.1962\,\text{W}$

$P \approx 0.20\,\text{W}$ **[1 mark]**

d) $P = \dfrac{V^2}{R} = \dfrac{9^2}{1000} = 0.08\,\text{W}$ **[1 mark]**

3. a)

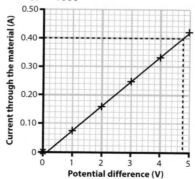

[3 marks: 1 mark for correct axes; 1 mark for points plotted correctly; 1 mark for straight line drawn correctly]

b) $\rho = \dfrac{RA}{l}$ and $V = IR \rightarrow R = \dfrac{V}{I}$

$\rho = \dfrac{V}{I} \cdot \dfrac{A}{l}$ **[1 mark]**

and rearranging gives $I = \dfrac{V}{\rho} \cdot \dfrac{A}{l}$

$\therefore I = \dfrac{A}{l\rho} V$ **[1 mark]**

$\dfrac{A}{l\rho}$ represents the slope (gradient) of an I–V graph, and

so $\dfrac{1}{R}$ **[1 mark]**

c) gradient, $m = \dfrac{0.4 - 0}{4.80 - 0.15} = 0.086$ **[1 mark]**

$m = \dfrac{A}{l\rho} \rightarrow \rho = \dfrac{A}{lm} = \dfrac{\pi\left(0.115 \times 10^{-3}\right)^2}{1.02 \times 0.086}$ **[2 marks]**

$\rho = 4.7 \times 10^{-7}\,\Omega\,\mathrm{m}$ **[1 mark]**

d) The material is metallic **[1 mark]** since ρ values for metals are typically in the range 10^{-7} to $10^{-8}\,\Omega\,\mathrm{m}$ **[1 mark]**

Emf and Internal Resistance
QUICK TEST (page 114)

1. The emf measures the electrical energy gained by each unit of charge that passes through the source
2. volts
3. because of the internal resistance of the cell/battery
4. This is the potential difference across the terminal of the source when it is in a circuit
5. by using a high-resistance voltmeter across the terminals of a cell and a variable resistor to change the current
6. internal resistance $(-r)$ and emf, respectively
7. the sum of the power delivered to the load resistance and the power wasted in the cell owing to the internal resistance
8. when the load resistance is equal to the internal resistance of the cell/battery
9. (i) the sum of the individual emf's (ii) the value of the emf of just one cell
10. (i) $15\,\mathrm{V} - 13.5\,\mathrm{V} = 1.5\,\mathrm{V}$ (ii) 'lost' volts (iii) $r = \dfrac{V}{I} = \dfrac{1.5}{0.3} = 5\,\Omega$

PRACTICE QUESTIONS (page 115)

1. a)

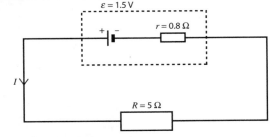

[3 marks: 1 mark for correctly drawn circuit; 1 mark for correct labels; 1 mark for correct current direction]

b) (i) $R + r = 5.8\,\Omega$ $I = \dfrac{V}{R+r} = \dfrac{1.5}{5.8} = 0.26$ A **[2 marks]**

(ii) $V = IR = 0.26 \times 5 = 1.3\,\mathrm{V}$ **[1 mark]**

(iii) $P = I^2R = 0.338 \approx 0.34\,\mathrm{W}$ **[1 mark]**

(iv) $P = I^2r = 0.054 \approx 0.05\,\mathrm{W}$ **[1 mark]**

2. a)

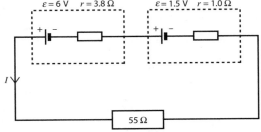

[2 marks: 1 mark for correctly drawn circuit; 1 mark for correct labels]

b) $R_{\text{total}} = 3.8\,\Omega + 1.0\,\Omega + 55\,\Omega = 59.8\,\Omega$

$R_{\text{total}} \approx 60\,\Omega$ **[1 mark]**

c) $V = IR$ $\therefore I = \dfrac{V}{R} = \dfrac{7.5}{59.8} = 0.125$ A **[2 marks]**

d) p.d across the $55\,\Omega$ resistor:

$V = IR = 0.125 \times 55$

$V = 6.897\ldots \approx 6.9\,\mathrm{V}$ **[2 marks]**

3. a) $\varepsilon = IR + Ir$

$IR = \varepsilon - Ir$

$V = \varepsilon - Ir$

$V = -rI + \varepsilon$ **[2 marks]**

"$y = mx + c$"

Therefore plot V (terminal potential difference) against I (current); the gradient is $-r$ (where r is the internal resistance) and the y-intercept gives ε (the emf of the cell) **[2 marks]**

b) (i) emf, $\varepsilon = 1.5\,\mathrm{V} \pm 0.1\,\mathrm{V}$ **[1 mark]**

(ii) gradient $= \dfrac{\Delta y}{\Delta x} = \dfrac{0.8 - 1.5}{1.0 - 0} = \dfrac{-0.7}{1} = -0.7$

gradient $= -r$

$\therefore$ internal resistance, $r = -(-0.7) = 0.7\,\Omega \pm 0.1\,\Omega$ **[2 marks]**

The Potential Divider and its Applications with Sensors
QUICK TEST (page 118)

1. a fixed or variable output potential difference or voltage
2. series
3. the same/identical, and half the source potential difference
4. $\dfrac{5}{10} = 0.5$
5. 2 V
6. sensor circuits
7. thermistors; light-dependent resistors (LDRs)
8. a transistor
9. **Any two from:** turning lights on; turning the heating system on; adjusting the volume of a stereo system

PRACTICE QUESTIONS (page 119)

1. a) As R_1 and R_2 are equal, the voltage is split equally between the two: $V = 6\,V$ **[1 mark]**

b) If $R_{var} = 8\,\Omega$ and $R_2 = 2\,\Omega$ then
$$V_{var} = 12\left(\frac{8}{10}\right) = 9.6\,V \text{ [2 marks]}$$

c) $R_{total} = 8 + 2 = 10\,\Omega$

using $V = IR$ and thus $I = \dfrac{V}{R}$:
$$I = \frac{12}{10} = 1.2\,A \text{ [2 marks]}$$

d) for two $6\,\Omega$ resistors in parallel, $R = 3\,\Omega$ **[1 mark]** and
$$\therefore R_{total} = 3 + 2 = 5\,\Omega \text{ [1 mark]}$$
$$I = \frac{12}{5} = 2.4\,A \text{ [1 mark]}, \text{ i.e. the current doubles}$$

2. a) At 24°C, $R_{thermistor} = 1.2\,k\Omega = 1200\,\Omega$

$V_1 = 2.2\,V$

$\therefore$ potential difference across thermistor $= 6.0 - 2.2$
$$= 3.8\,V \text{ [1 mark]}$$

b) $\dfrac{V_{thermistor}}{V_1} = \dfrac{R_{thermistor}}{R_1} \Rightarrow R_1 = R_{thermistor}\left(\dfrac{V_1}{V_{thermistor}}\right)$

$\therefore R_1 = 1200\left(\dfrac{2.2}{3.8}\right) = 694.7 \approx 695\,\Omega$ [accept $700\,\Omega$]

[2 marks]

c) $V_{thermistor} = IR_{thermistor} \rightarrow I = \dfrac{V_{thermistor}}{R_{thermistor}}$

$$I = \frac{3.8}{1200} = 0.00316\,A \approx 0.32\,mA \text{ [2 marks]}$$

d) temperature increase $\rightarrow R_{thermistor}$ decreases **[1 mark]**

e) Voltage across resistor would increase since it now has a greater proportion of the circuit resistance across it according to
$$V_1 = V_{thermistor}\frac{R_1}{R_{thermistor}} \text{ [2 marks]}$$

3. a)

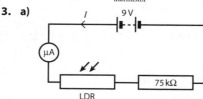

LDR

[3 marks: 1 mark for correctly drawing circuit; 1 mark for correct labels; 1 mark for correct components]

b) dark $\rightarrow R_{LDR} = 180\,k\Omega = 180 \times 10^3\,\Omega$

$R_{total} = \left(180 \times 10^3\right) + \left(75 \times 10^3\right)\Omega$
$$= 255 \times 10^3\,\Omega \text{ [1 mark]}$$

using $V = IR_{total} \rightarrow I = \dfrac{V}{R_{total}} = \dfrac{9}{255 \times 10^3}$ **[1 mark]**

$I = 35.3 \times 10^{-6}\,A = 35\,\mu A$ **[1 mark]**

c) $V_{LDR} = IR_{LDR} = 35 \times 10^{-6} \times 180 \times 10^3$
$$= 6.35\,V \approx 6.4\,V \text{ [2 marks]}$$

d) light $\rightarrow R_{LDR} = 10\,k\Omega = 10 \times 10^3\,\Omega$

$R_{total} = 85\,k\Omega = 85 \times 10^3\,\Omega$ **[1 mark]**

$I = \dfrac{V}{R_{total}} = \dfrac{9}{85 \times 10^3} = 105.9 \times 10^{-6}\,A$

$I = 106\,\mu A$ **[1 mark]**

and $V_{LDR} = IR_{LDR} = 106 \times 10^{-6} \times 10 \times 10^3$
$$= 1.06\,V \text{ [1 mark]}$$

e) if $V_{LDR} = 3\,V$ then $V_R = 6\,V$

$\dfrac{V_{LDR}}{V_R} = \dfrac{R_{LDR}}{R} \rightarrow R_{LDR} = R\left(\dfrac{V_{LDR}}{V_R}\right) = 75 \times 10^3 \times \left(\dfrac{3}{6}\right)$

$R_{LDR} = 37.5\,k\Omega$ **[2 marks]**

Index

Acknowledgements

The author and publisher are grateful to the copyright holders for permission to use quoted materials and images.

Cover & P1: © Shutterstock.com/Yellowj

P48 © sciencephotos / Alamy Stock Photo
All other images are © Shutterstock.com and © HarperCollinsPublishers Ltd

Every effort has been made to trace copyright holders and obtain their permission for the use of copyright material. The author and publisher will gladly receive information enabling them to rectify any error or omission in subsequent editions. All facts are correct at time of going to press.

Published by Letts Educational
An imprint of HarperCollinsPublishers
1 London Bridge Street
London SE1 9GF

ISBN: 9780008179106

First published 2016

10 9 8 7 6 5 4 3 2 1

© HarperCollinsPublishers Limited 2016

British Library Cataloguing in Publication Data.
A CIP record of this book is available from the British Library.

Series Concept and Development: Emily Linnett and Katherine Wilkinson
Commissioning and Series Editor: Chantal Addy
Author: Ron Holt
Project Manager and Editorial: Andrew Welsh
Cover Design: Paul Oates
Inside Concept Design: Paul Oates and Ian Wrigley
Index: Indexing Specialists (UK) Ltd
Text Design, Layout and Artwork: Q2A Media
Production: Lyndsey Rogers and Paul Harding
Printed in Italy by Grafica Veneta SpA

MIX
Paper from
responsible sources
FSC™ C007454